LITTLE SPECIES BIG MYSTERY

The story of
Homo floresiensis

LITTLE SPECIES BIG MYSTERY

Debbie Argue

MELBOURNE
UNIVERSITY
PRESS

MELBOURNE UNIVERSITY PRESS
An imprint of Melbourne University Publishing Limited
Level 1, 715 Swanston Street, Carlton, Victoria 3053, Australia
mup-contact@unimelb.edu.au
www.mup.com.au

First published 2022

Cover design by Pfisterer + Freeman
Typeset by Megan Ellis
Cover image by Paul A. Souders, courtesy Getty Images
Printed in China by 1010 Printing International Limited

A catalogue record for this book is available from the National Library of Australia

9780522877915 (paperback)
9780522879162 (paperback, signed)
9780522877922 (ebook)

To the memory of Emeritus Professor Colin Groves, eminent scholar and much valued teacher, colleague and friend.

24 June 1942 – 30 November 2017

Image provided by Debbie Argue.

Contents

Prologue

News of the discovery of a new species, *Homo floresiensis*, burst upon an unsuspecting world in 2004: a series of small, human-like bones had been discovered during archaeological excavations in a cave on the Indonesian island of Flores. In one swoop, much of what we thought we knew about human evolution was challenged. Could there really have been a population of tiny beings around 1 metre tall *and* living at the same time as we *Homo sapiens*? Such a scenario could hardly have been envisioned before this remarkable discovery was made. The species, with its odd combination of a small head, short legs and elongated feet, captured the public's imagination and the nickname 'the Hobbit' was quickly adopted—a reference, of course, to JRR Tolkien's well-known book of the same name.

In science, new ideas are subject to examination and questioning, and in this case there was swift and sustained controversy, with the bones interpreted by some to be nothing more than the remains of diseased modern humans. Meanwhile, those who recognised that this discovery did indeed herald something new wanted to know just where this enigmatic species would fit on the human evolutionary tree. Who were its ancestors? And how did this species get to an island that had never been attached to a mainland?

This book is a personal narrative about the discovery of *H. floresiensis* and all that happened afterwards—the controversies, the puzzles, the questions and the answers. The story is told from the perspective of experts worldwide who have conducted groundbreaking research on the characteristics and evolution of *H. floresiensis*. I am most fortunate to have been closely involved in the study of this intriguing and perplexing species since it was first announced in the scientific literature. I still have a sense of wonder that my life took this trajectory: never had I dreamt that such great excitement awaited me, and yet, looking back on it now, I can see that the past had held me in its sway since I was quite young.

It is all so distant from my eleven-year-old self who loved exploring the ruins of old Australian homesteads, inspired by a piece in a magazine for schoolchildren provided by the NSW Education Department. That story began when two children who had recently arrived in a small country town from faraway England were wandering along a street, feeling disillusioned with their new circumstances. Sensing their loneliness and frustration, a woman, leaning over her garden gate, suggested they go explore along a country road where they would find an old house. Pushing through matted grass and an overgrown garden, the children found the abandoned house. It was nothing like they'd ever seen. Gingerly the children entered. They tried to understand what each room would have been used for, imagining who might have lived there all those years ago. What would life have been like back then? Later, they even returned with pencils and paper and drew up a floor plan to better understand the layout of the rooms. After using clues from elements of the house, and with the woman's help, the children worked out part of its history.

That it was possible to figure out something like this captivated me. From then on, whenever my family drove past abandoned old houses or ruins, I would badger my parents to stop the car. They would indulge my new interest (well, not always) and off we'd go, clambering over fences and through paddocks.

It was at around this age that I read *The Rocks of Honey* by Patricia Wrightson. The novel follows a boy whose family has recently moved to the Blue Mountains, west of Sydney. He befriends an Aboriginal boy as well as a girl who's also new to the area. When an Aboriginal Elder tells them the legend of the nearby Rocks of Honey, guardian of a mysterious stone axe, the children set out to find the tool. They succeed but then have to make a choice about the right thing to do. This was my introduction to aspects of Aboriginal culture, and stone tools. How I wanted to learn more about the people from this culture who once made these tools![1]

These two narratives tucked themselves away in the back of my mind to percolate over the years. Following secondary school, I gained admission to the Australian National University (ANU) in my hometown of Canberra, where I studied geography (great fieldwork), economics (challenging and interesting), history ('You'll never pass this subject,' my helpful lecturer told me)[2] and Asian civilisation (archaeology included). Then it was off to Teachers' College in Sydney: flatting, parties, new friends. The following year I was out teaching at a secondary school.

But what I really wanted to do was travel. By 1974 I'd earned enough money to fund a trip and, together with a friend, I sailed out of Sydney on the ship *Britanus*. We docked at Portsea and set off backpacking around the United Kingdom and mainland Europe. My love of archaeology re-emerged: Stonehenge, Skara Brae, the Roman aqueducts in Italy—all so different from the ruins I'd come to know back home. Six months later, not wanting my adventures to end, I signed on for a bus trip from London to Kathmandu. But while we were in Beirut, our travel company went bankrupt. We were stranded. Our ongoing-travel funds had been swallowed up and, poignantly, our bus had to be abandoned on the street. Left to our own devices, many of us opted to forge ahead together, using whatever public transport was available: truck convoys across the desert, dhows across the Persian Gulf. Our journey was peppered with the archaeological sites I'd learned about while studying ancient history: Baalbek, Persepolis, Petra. We were exposed to cultures that seemed so exotic. And there was romance in the air: it was on this trip that I met my future husband, Fraser.

Life and parenting in New Zealand and Canberra eclipsed the next decade or so. It was not until our children were older that I had the opportunity to re-engage with archaeology. I joined the Canberra Archaeology Society and volunteered for every dig I could. I went back to university and enrolled in another Bachelor of Arts degree, focusing on prehistory and archaeology. I not only passed my coursework, I got good grades, much better than in my first degree. But one subject had me stumped: human evolution. I failed the first-semester test. Disheartened but not beaten, I knuckled down and got to work on this difficult (for me) topic, achieving a high distinction at the end of the year—and a new respect for the subject.

Over the next few years I worked as an archaeologist, but I found myself reading books on human evolution in my spare time. I was absorbed, even though my reading was unstructured, all over the place, and I had no clear goal in mind. One day I ran into my human evolution lecturer, Professor Colin Groves, and mentioned my difficulties to him. He suggested I focus on just one hominin species to begin with, and he kindly offered me access to his personal library, which was open to all students. It was on one of my many visits there that he casually suggested: 'You could do your PhD, you know.' And so I did, starting just months before the incredible announcement of *H. floresiensis*. Little did I know what a profound influence this little being would have on the direction of my academic life.

I had the good fortune to collaborate with Colin over many years, working on *H. floresiensis*, and he features prominently in this book. He was one of the few taxonomists whose knowledge spanned primates, mammals and the evolution of hominins. Here I provide some insight into Colin's life and personality, knowing that just a few paragraphs cannot hope to do justice to this extraordinary scholar.

In 2006, the internationally acclaimed primatologist Jane Goodall was the guest speaker at an event in Canberra. She began with: 'I like Canberra, and you do have a significant advantage. You've the best taxonomist—*the* taxonomist—the remarkable Colin Groves.' Colin once mentioned that he became fascinated with wildlife after his grandfather gave him a book about animals. At one of the many get-togethers Colin and his wife Phyll hosted at their home, Colin showed us the very book. It was a lovely edition, beautifully illustrated, and we could see how he had become so inspired.

Over his career Colin described sixty-two new species, one of which was a new hominin, *Homo ergaster*, that he and his colleague Vratislav Mazák identified from a 1.5-million-year-old jawbone in 1975. The other new species that Colin named included pigs, deer, gazelles, duikers, gibbons, tarsiers, lemurs, monkeys, civets, possums and mosaic-tailed rats.[3]

So impressed was I with Colin's prodigious knowledge of the natural world, I once asked him if there was any group of animals that was not a focus of research for him. He had to think for quite a while before responding, somewhat tentatively: 'Maybe fish.'

In 1989 Colin published *A Theory of Human and Primate Evolution*, a go-to book that sorts out what happened in human evolution—true to form, everything Colin espouses in this book is based on data analysis and is testable scientifically. Colin's publications number at least 775, including books, chapters, scientific and popular articles, book reviews, obituaries, and newspaper letters and articles.[4] The very first species he named, as a new subspecies, remains the largest living mammal described in recent generations—the Bornean rhino, critically endangered and very nearly extinct. Colin was working right up until shortly before he passed away. In 2017 he contributed to the identification of a new subspecies of orangutan: Tapanuli orangutan (*Pongo tapanuliensis*) of Sumatra, one of only eight living great apes on our planet.[5] Colin also co-authored the announcement of a new tarsier species, *Tarsius niemitzi sp. nov.*, which was found in the Togean Islands of central Sulawesi—the paper was published posthumously, in 2019.[6]

Of abiding importance to Colin was the conservation of wildlife and their habitats; often speaking at fundraising events, he was never shy about voicing his opinion on controversial topics.[7] So dedicated was he to this cause that he quietly and generously set up a Primate Conservation Grant at the ANU, out of his personal funds, to support graduates in their research endeavours.

Colin had a unique way of engaging people in science. I recall one occasion when he was speaking about primates during National Science Week in Canberra. Everything starts off in a normal way. He talks about gorillas, while wearing a T-shirt with a picture of a gorilla on it. Then a collective gasp goes around the room—Colin is taking off his T-shirt! But it's OK. He's wearing another T-shirt underneath, one with an image of the primate he is about to discuss. Soon he strips off that T-shirt, then a third, and a fourth. The audience is in fits of laughter; the children can hardly contain themselves. Seven or eight T-shirts later, to our great relief, he does not remove the final garment to reveal a modern human primate's chest.

Students came to the ANU from around Australia, Indonesia, Malaysia, Vietnam and China to study primatology, human evolution and mammals under Colin's tutelage. In turn, many now contribute to their home country's pool of conservation, research and teaching experts. Over the years, several such students have quietly told me that they came to the School of Archaeology and Anthropology because Colin not only replied to their email of inquiry but did so in a very encouraging way. Most colleagues will remember Colin's 'open door' policy, how at any time individuals could turn up to have a chat about, or seek help with, their work, or simply ask a question. The longest I recall waiting for Colin's attention was approximately three minutes, while he finished a task on his computer.

Colin's extensive library was student heaven. Having extended along three walls of his office, still it grew, and an extra floor-to-ceiling bookcase had to be erected in his already book-crammed room. He knew where almost every book was on his shelves, and he seemed to know which of his thousands of reprints and photocopied articles would have just the information a student needed. Colin also recognised that students could become isolated in their rooms, beavering away (one hopes) at their research. He therefore set up a bioanthropology lunch group, where once a week we'd walk over to a nearby café and have conversations that ranged over anything and everything.

The tearoom at the School of Archaeology and Anthropology was itself a lunchtime institution. Colin would come along and sit in his favourite chair and we'd talk about current events, academic questions, books, movies, music and TV shows (he was an avid fan of the UK quiz show *QI*). If any visiting scholars were around, Colin would bring them along too, giving us a chance to interact with experts we might not normally have met.

A colleague, Dr Christine Cave, recalls:

> Colin enjoyed the Harry Potter books. One day in the tearoom, the conversation turned to Harry Potter and bezoars, and bezoar goats. In the Potter books, bezoars are stones taken from the stomachs of goats which will counteract the effects of any poison. Colin was able to tell us that, indeed, there are such things as bezoar goats or ibex, living in

> Eastern Europe, Turkey and Iran. They indeed do swallow stones, and legends of the people of these areas state that these bezoar stones have medicinal properties.[8]

It was also important to Colin to counter pseudoscience, which utilises statements, beliefs and practices that claim to be both scientific and factual but are incompatible with the scientific method; examples include dowsing and horoscope predictions. In refuting a particular aspect of pseudoscience, Colin would be rational and courteous, presenting the relevant facts in a straightforward manner. He was an active member of the Canberra Skeptics, awarded Honorary Life Membership of the group in 2006.

Cryptozoology, the study of the possible existence of as-yet-unknown animals, is a field Colin took seriously. It's not that he 'believed in' any particular thing. He merely espoused a rigorous scientific approach: claims had to be testable. As far as I know, no claim of the existence of a heretofore unknown animal survived his scrutiny, yet he retained an open mind about such matters, even subscribing to journals on this subject.

By now it will be clear that one of the many things I admired about Colin was that new ideas and different ways of thinking did not faze him so long as they were based on a scientific approach that could be tested by other researchers. He was a clear and objective thinker, a skill he imparted to students. This, and his gentle nature, meant that students felt confident in floating ideas by him. The worst that could happen was a querying 'Are you *sure*?', in which case you knew for certain you were on quite the wrong track and a rethink was in order.

Colin was awarded a BSc in Anthropology in London in 1963 and received his PhD three years later. He was then invited to apply for a lectureship at the Department of Archaeology and Anthropology at the ANU, arriving in Australia in 1974. In 1980 he was appointed Senior Lecturer, and in 1988 Reader. In 2000 he was awarded a professorship, and upon his retirement in 2016 he was bestowed with an emeritus position. In recognition of his contribution to scholarship, Colin was admitted to membership of the New York Academy of Sciences (1995), elected a Fellow of the Australian Academy of the Humanities (1998), awarded Honorary Life Membership of the American Society of Mammologists (2013), given a Conservation International Award for Primate Conservation (2014), and presented with a Lifetime Achievement Award (2016) from the International Primate Society.

Colin enriched my life, as he did for so many others. I feel privileged to have had him as supervisor for my Master of Arts and PhD degrees, and to have worked with him on the *Homo floresiensis* question.

Timeline of species

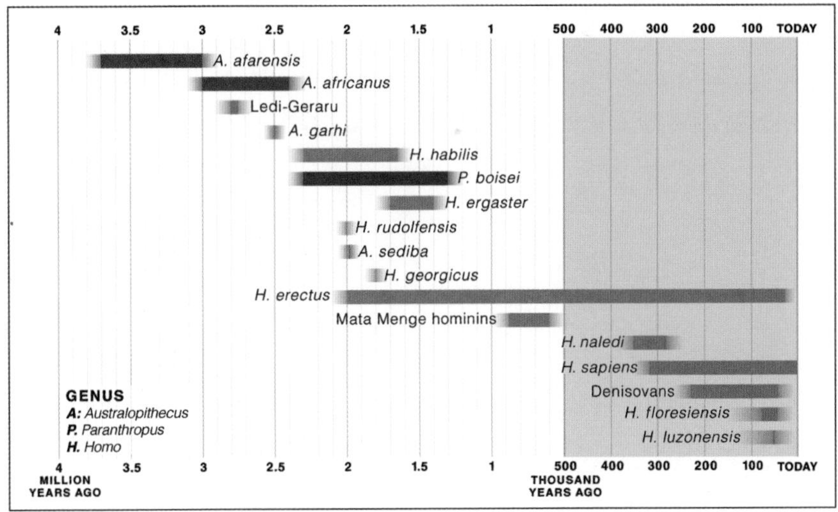

A timeline of the species discussed in this book; note that this diagram does not include all species known in human evolution. For stories about some of the discoveries of hominin fossils, see Appendix B. Prepared by Geraldine Cave.

1

The discovery

In late 2003, at a conference on Australian archaeology at Jindabyne in New South Wales, I met a couple of students who told me that something amazing was coming out of Indonesia, but they couldn't say any more than that. Interesting? Tantalising? Yes! I waited months for some kind of evolution bombshell to be unveiled in *Nature*, which surely was where such a dramatic find would be revealed. Time went by and still nothing. Had I read too much into what the students had said?

Then, on 28 October 2004, there it was, amid a torrent of publicity: a new species of hominin named *Homo floresiensis*. This incredible discovery, one that would rattle the status quo on human evolution, comprised a series of bones excavated in Liang Bua cave on the island of Flores. The breakthrough was made by a team of Indonesian and Australian researchers led by Professor Mike Morwood, then at the University of New England, and Thomas Sutikna from the Jakarta-based Indonesian National Research Centre for Archaeology (ARKENAS).

As I read the publication by palaeoanthropologist Professor Peter Brown (University of New England) and colleagues that described bones representing archaic-looking individuals 1 metre tall who lived until a mere 18 000 years ago,[1] my reaction was probably much the same as everyone else's: 'How could this be?' but also 'How wonderful, what a mystery!'

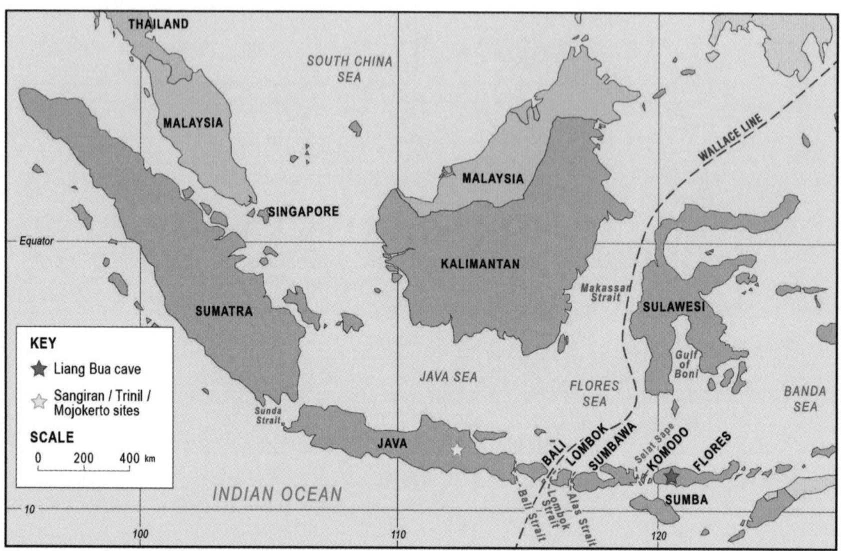

Map of the Indonesian region: the red star indicates the location of Liang Bua cave; the yellow star indicates the Sangiran region of Java. Prepared by Geraldine Cave.

The discovery challenged much of what we thought we knew about human evolution. What was astonishing was that we now had physical evidence of tiny, upright-walking human-like beings that looked like something we'd expect from African hominin sites of two million years ago, yet these individuals until very recently were living half a world away from that continent. And could there really have been a population of diminutive beings that lived at the same time as us? Such a scenario seemed impossible, yet here was evidence of it.

At the time, I was studying one-million to two-million-year-old fossil hominin skulls from Africa for my PhD. Even though the newly discovered bones seemed to date from much more recent times, I was surprised at the similarity between the *H. floresiensis* skull and the much older African skulls. Somewhat tentatively, I asked my PhD supervisor, Professor Colin Groves, if I could include this fossil in my study—despite it potentially being over a million years *younger* than my target fossils. Colin reflected for a few seconds and then said, 'Yes, why not.' And so began my fascination with this enigmatic species, and my long collaboration with Colin in the world of human evolution.

How did archaeologists come to be digging in that particular cave on Flores? And what had they expected to find? To find out, we need to go back to an evening in 1995, when then Dr Mike Morwood, his colleague Doug Hobbs and Wunambal Aboriginal Elder Jack Karadada were in the Kimberley region of north-western Australia, reflecting on their many seasons of studying the local Aboriginal rock art. As they sat at their camp spot, watching the sun drop into the sea and with a full moon rising, they talked about setting up an ambitious archaeological project to search for the origins of the first Australians.[2] Because the fossil bones of an extinct member of the elephant family, the stegodont, had been found at an archaeological site called Mata Menge on Flores and dated to 700 000 years ago, Mike and Professor Bert Roberts (University of Wollongong) had reasoned that evidence for much later modern humans could be discovered in Indonesia. But Indonesia is a country with thousands of islands, and the evidence could be on any one of them. So where would such a project begin?

In 1999, Mike, Doug and their Indonesian colleagues Fachroel Aziz (from the Geological Survey Institute, Bandung) and Jatmiko (ARKENAS) visited a large limestone cave in western Flores called Liang Bua. The cave was already a site of interest to Indonesians and others. Father Theo Verhoeven, a Dutch priest who had taught school classes in the cave in the 1940s, had been very keen on archaeology. He had excavated parts of the cave in the 1950s and 1960s, uncovering stone artefacts, pottery, and six burial sites of modern humans along with grave goods. The cave again became a focus of archaeological work when Professor Soejono, the then Director of ARKENAS, undertook a long-term excavation project there between 1978 and 1989, during which he discovered evidence of 10 000 years of modern human occupation.[3]

When Mike Morwood saw the cave, he was particularly impressed by its size—50 metres wide and 30 metres high—and its feeling of spaciousness, with its well-lit northern outlook and its flat, dry clay floor. Although Professor Soejono had excavated to at least 3 metres, no-one knew how deep the cave sediments could be. However, an adjacent cave, from which most of the sediments had been washed out, showed that Liang Bua might extend to quite a depth. Mike surmised that the cave could well reveal archaeological evidence concerning when modern humans first arrived in Indonesia. He viewed this cave's potential as 'the best I had seen in more than 30 years'.[4]

This enticing possibility was the catalyst for Mike and a team from ARKENAS, including Professor Soejono, Thomas Sutikna, Rokus Awe Due and Sri Wasisto, with the assistance of twenty local Manggarai workers, to begin test excavations in 2001. As work proceeded, the team came up against a challenge reminiscent of Professor Soejono's experiences of excavation in the 1980s. Soejono's team had hit a thick, well-cemented layer of material. Not unnaturally, they thought this represented the end of the cultural sequence, meaning that there would be no evidence of human activity below this layer. Mike and Thomas faced a similar situation. Thomas recalls:

> We were down 3.5–4 metres and were faced with a massive limestone block cemented by flowstone with a big stalagmite in the middle of it. Mike and I thought we just *had* to get below this. I chopped at the edge of the flowstone and found not solid block, but sediment![5]

And so on went the digging. The first surprise came when they found that the layers of sediment were packed with stone tools—'5000 artefacts per cubic metre of deposit! Here was human activity by the bucketload,' writes Mike Morwood.[6] No-one knew who could have made these tools. Then, an even more astounding discovery: among the tools were high concentrations of teeth (mostly) of stegodonts. Also found were some human-like bones.

There was such good potential for further archaeological work that Mike Morwood and Bert Roberts applied for an Australian Research Council (ARC) grant. Their application, entitled 'Astride the Wallace Line: 1.5 Million Years of Human Evolution, Dispersal, Culture and Environmental Change in Indonesia', was a somewhat wider-ranging topic than Mike, Jack and Doug had imagined as they'd watched the sun set over the Kimberley back in 1995. But times had moved on. Bert Roberts and his colleagues' work at Malakunanja cave (now known as Madjedbebe) in Arnhem Land, in the Northern Territory, had shown that Aboriginal people were on the continent between 50 000 and 60 000 years ago, at least 10 000 years earlier than had been thought.[7] 'And Carl Swisher and colleagues found that *Homo erectus* survived on Java until 40,000 to 50,000 years ago,' says Bert, 'raising the tantalising possibility that moderns were around at the same time as *H. erectus* on Java. We wanted to look for evidence of any interaction between these species.'[8] A tantalising prospect indeed.

These exciting questions did the trick and the ARC granted funding in October 2002.[9] But even before the grant money arrived, the enthusiastic team was back in Liang Bua cave, courtesy of some research funding of Bert Robert's.[10] For this new work, Professor Soejono, Mike and Jatmiko assembled a larger team of archaeologists: Thomas Sutikna, Wahyu Saptomo, Sri Wasisto, Rokus Awe Due and Doug Hobbs, plus the specialists Fachroel Aziz (palaeontology), Carol Leftner (University of Wollongong; plant remains), Dr Gert van den Bergh (who had earlier assisted with the stegodont remains) and Dr Mark Moore (University of New England; stone artefact expert). Bert's PhD student Kira Westaway (Macquarie University[11]) joined the team to establish the environmental context for the site, and dating experts Paul O'Sullivan (GeoSep Services, USA) and Chris Turney (University of New South Wales) rounded out the dating team.[12]

The Liang Bua team must have been impressed by their discovery of stegodont bones and all those tools, but they were in for a bigger surprise. Further down, at 6 metres, the archaeologists came upon a human-looking lower-arm bone. Thomas recalls: 'I was so surprised. This arm bone was so small, and its shape was strange—it was not right for a human arm bone, just not right.'[13] Although the arm bone was from an adult, it was very short and it had an odd curvature in it. The team could see that this was no modern human, but none of them had the anatomical expertise to identify its species. It was only when Professor Hisao Baba of the Natural Science Museum in Tokyo compared a cast of the arm bone with a large collection of bones at the museum, that he could declare it was definitely human, although which kind of human was not known. The team may not have known it at the time, but they had discovered the first bone of a remarkable new species of hominin that, three years later, would be named *Homo floresiensis.* This single, small, enigmatic arm bone must rank as one of the most significant discoveries in human evolution. Yet it would remain unpublished and unheralded, and be eclipsed by a spectacular discovery two years later by the same team in the same cave.

With an unusually shaped arm bone, a stack of stegodont teeth and a plethora of stone tools, dating the strata was now crucial. Bert, along with Jack Rink, a dating specialist from McMaster University, Canada who was visiting the University of Wollongong at the time, were on site within a week. This was no small feat considering the amount of organisation that is required to source and transport bulky scientific equipment internationally. Both researchers are experts in the highly impressive-sounding dating techniques of thermoluminescence (TL), optically stimulated luminescence (OSL) and electron spin resonance (ESR) dating. In simple terms, TL can tell you when the grains in anything were last heated, so this is good for dating, for example, volcanic material; OSL dates the last time a quartz grain in sediment saw sunshine; and ESR can be used to date bones and tooth enamel, among other things. The preliminary dates obtained by Jack Rink for the stegodont molars showed that the team was working within a 100 000-year sequence that had enclosed the mysterious human-like bones. Who was this person? Why were the stegodont remains so recent compared to the 700 000-year-old remnants at Mata Menge? Furthermore, these animals did not normally frequent caves, so why were they there?

The archaeologists hit pay dirt the following year with the discovery of a small partial skeleton, with some of the bones still attached to others—we call this 'partially articulated'. Jatmiko remembers the moment:

> Wahyu [Saptomo] called me to come down into the excavation hole because he saw Benyamin [Tarus] finding some bones of a human skull. Wahyu didn't know the condition of the in situ skeleton, so Benyamin and I made it clean [in archaeology, this means carefully brushing soil from the find] and exposed the bones.[14]

Jatmiko was very surprised because as they worked he could see that they were uncovering a partial skeleton. He and Wahyu took a comprehensive set of photos of the bones as found before going back to base camp—Hotel Sinda, in Ruteng, not far away. 'We showed the photos to Thomas and Rokus,' he says. 'The next day Thomas and Rokus make good preparation of the skeleton and then it was extracted and removed from the excavation hole.'[15]

The final bone tally of the partial skeleton included its skull, jaw, arm and leg bones, shoulder bones, and parts of the pelvis, hands, feet, ribs and vertebrae. But it was so different from us. The skull, with very thick bones, had a sloping forehead and no chin. Rokus recognised immediately that this was not a modern human child. As he explains,

> when they found the skull they called me and I went into the hole [the excavation]. I first thought it was a man's skull but then saw it was so small, like a child's. Maybe a child of ten years. We brought it to the hotel to clean it. I realised it still had its teeth but they were the teeth of an adult who was 28 or 30 years old. That day, all five of us were delighted and surprised by that discovery.[16]

When the researchers carefully removed the soil from the jaw of the newly discovered skeleton, now named Liang Bua 1 (LB1), they saw that not only was the dentition fully adult but the teeth were very worn down. Even though the bones represented a person who was just over 1 metre tall, it was a mature adult when it died. This was the smallest fossil member of the genus *Homo* known at the time.

We do not have to call this individual 'it', however, because later, when Peter Brown and colleagues studied the pelvis, they could see that it flared widely and they considered that LB1 was probably female.[17] Although it is not known how she died, the archaeological evidence shows that she had not been deliberately buried but rather, after death, she had sunk into mud in a shallow pool of water where she was slowly covered by silt.

Further excavation showed that LB1 was not alone. There were bones from a number of adults and children, including another adult jaw. The bones of all the adults represented very small individuals. None were the result of deliberate burial and none were of modern humans. LB1 was represented by the most bones, but another individual, labelled LB6, had quite a few too, with arm bones, a shoulder bone, six fragments of finger bones, and a toe bone remaining. Of the other individuals, however, not much remained.

Top: Liang Bua cave in 2009, with excavations well underway. LB1 had been discovered in excavations next to the pits being dug on the right-hand side of the cave, roughly where the timber planks are resting. After excavation, the pits are back-filled for preservation, which is why they were not visible when this photo was taken. Image provided by Debbie Argue.

Bottom: The discovery of the skeleton—Jatmiko cleans up the surface of the bones in the excavation. Image provided by Jatmiko.

Two were represented by an arm bone, one by a wrist bone, one by a foot bone, and one by a single finger bone. A knee bone, an arm and a leg bone, and fragments of pelvis and ribs, represented three more individuals.[18] There was, then, a marked difference in the number and types of bone for these individuals compared to the partial skeleton LB1—just why this is so has yet to be determined. Professor Bill Jungers (from Stony Brook University in New York State) and colleagues described all the parts of the lower skeleton, concurring that LB1 was probably female, while noting that she appears to be the largest individual recovered for *H. floresiensis*.[19]

When the bones of the diminutive human were first discovered, the team simply had no idea to which species it belonged. It was necessary to compare the bones with those of *H. sapiens* and archaic hominin species to see if they were similar to anything already known. Mike called in Dr Peter Brown from the University of New England.[20] Peter studied the LB1 bones in relation to those of the three-million-year-old species *Australopithecus africanus*; the 1.8 million-year-old *Homo georgicus* from the Caucasus country of Georgia; the 1.5–1.8-million-year-old *H. erectus* from Java; and the 1.5-million-year-old *H. ergaster* from Africa.[21] Peter saw that LB1 showed a mix of archaic and modern characteristics, and that her bones and those of the others from the Liang Bua excavation were unlike any other species of hominin. This meant that LB1 and the other individuals could now be declared a new species of *Homo*, to be known as *Homo floresiensis*.[22] She even acquired a nickname, the *Little Lady of Flores* or *Flo*—and later, affectionately, the Hobbit, from Tolkien's famous story.

Because we now have a partial skeleton and body parts from other individuals of *H. floresiensis*, we can build up a picture of what the species looked like. The discovery shows that the individuals were indeed small. LB1's adult stature was 106 centimetres (3 feet 6 inches). This is smaller than any modern human population. *H. floresiensis* is, in fact, more similar to chimpanzees in height. For her short stature, though, she had a large body mass distributed over her skeleton. Bill Jungers and Assistant Professor Karen Baab from Stony Brook University worked out that the body mass index for LB1 fell way outside the range of small-bodied modern humans—this little species was far stockier than any of us.

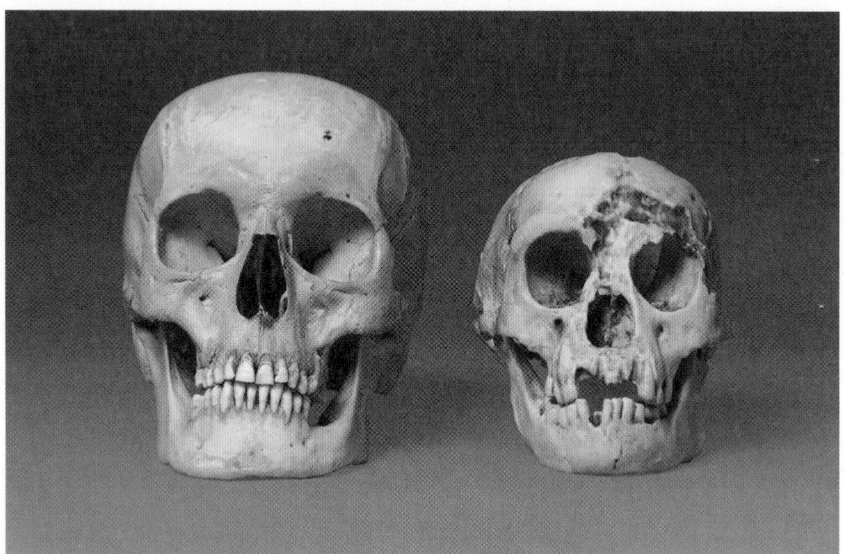

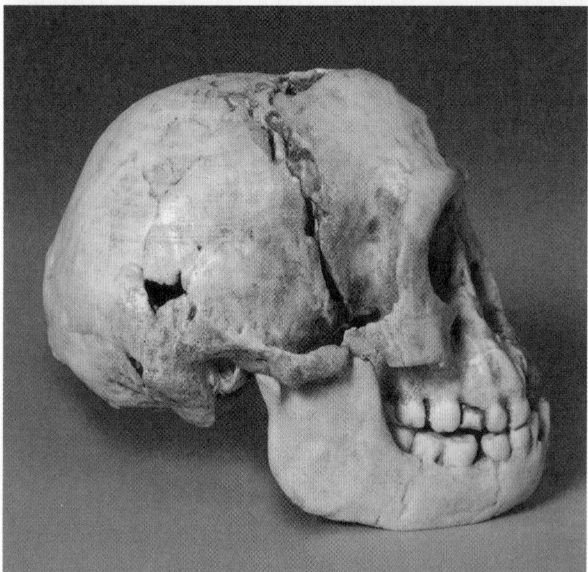

Top left, a modern human skull, and **right**, the skull of LB1 (replica). Images provided by Maggie Otto.

Bottom: LB1's prognathic face. Also note the mound of bone around the eye, the low form of the skull and the backward-sloping jaw. Image provided by Maggie Otto.

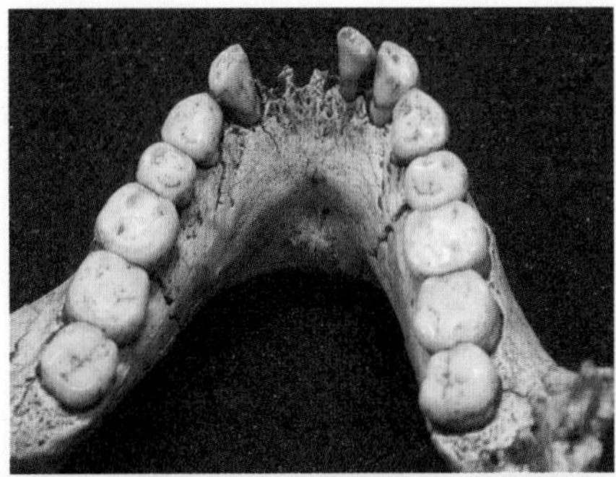
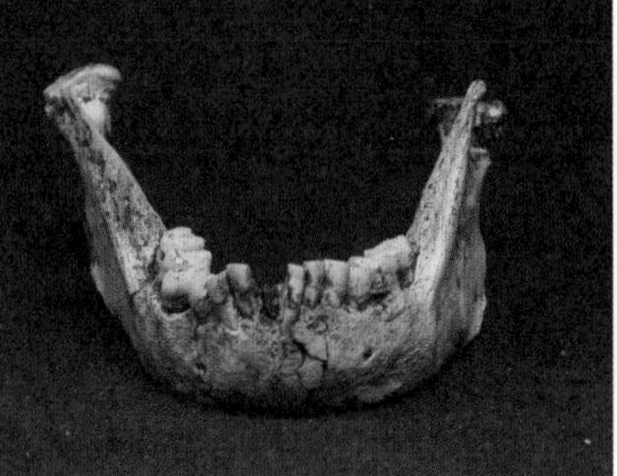
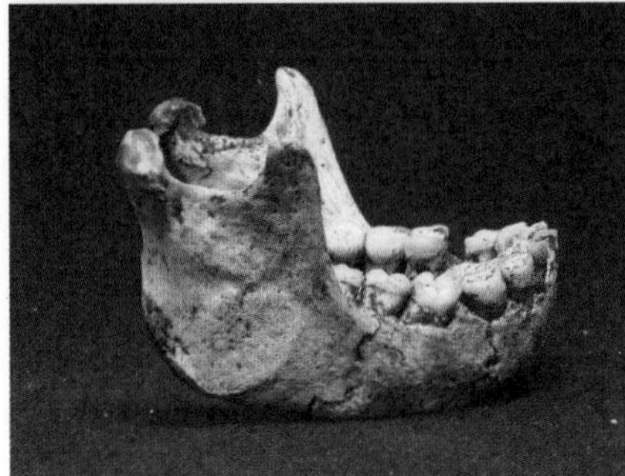

Views of LB1's jaw: **top left**, the internal buttressing; **right**, the smooth front of the jaw; and **lower left**, the receding front of the jaw. Images provided by Mike Morwood.

Her skull cavity is small, even for such a short being, with a capacity of only 426 cubic centimetres,[23] whereas ours average between 1300 and 1500 cubic centimetres. A mound of bone frames the upper and side regions of the eyes, and there is another mound extending up from each eyetooth (called canine juga). The forehead slopes back and the skull is low, with its widest part below the level of the ears; in comparison, the widest part of our skull is high, accommodating our modern expanded brain. In side view, LB1's face slopes outwards as it extends down from below the eyes. We call this condition prognathism. And the eye sockets look quite large.

LB1's jaw and a similar jaw, that of LB6, discovered in another part of the excavation, show that these individuals lacked a chin. Instead, this part of the jaw slopes back, and on the inside at the front is a short sloping ledge. Below that are two horizontal bony bars. These internal structures are ape-like and are the reinforcing parts of the jaw. Similar structures are seen on very early members of our genus, *Homo*, and on some other early hominin species, such as the australopithecines. They never occur on modern humans. Instead, our jaws are reinforced by bone build-up at the front. Because the jaw forms in utero, we know that the form of this part of the jaw is inherited, rather than being influenced by, for example, chewing on tough food.

The wrist bones are also ape-like. Associate Professor Matt Tocheri (Canada Research Chair in Human Origins at Lakehead University) is an expert on hand and foot bones (he is also an awesome jazz pianist). He has a special interest in the small bones of the hand, the carpals. He's studied hundreds of wrist bones of humans, chimps, gorillas, orangutans and baboons. His findings concerning LB1's wrist bones astounded us all—they were similar not to ours, but to those of apes.[24] As he explains:

> Basically it has a lot to do with one particular bone in the wrist—the trapezoid. It's the wrist bone that sits right underneath your index finger. In African apes and other nonhuman primates it's shaped like a pyramid but in modern humans and Neanderthals it is shaped like a boot. In *H. floresiensis* the trapezoid is a dead-ringer for what we see in apes and what would have been the ancestral condition for hominins. No modern human has this wrist.[25]

Later, when Colin Groves, Professor Bill Jungers and I were studying the *H. floresiensis* bones at ARKENAS, Bill echoed Matt's finding but added that the species could be partly arboreal; that is, it could have spent part of its time in trees, perhaps seeking food.

Although *H. floresiensis* individuals were small and, in respect of many features, unlike modern humans, they walked upright. The opening where the spine enters the skull—the foramen magnum—is positioned in the same place as seen in *H. sapiens*. It is unlikely, however, that this minuscule human walked in the way that we do. We learn from Bill Jungers and colleagues that *H. floresiensis* had relatively short legs and long flat feet. The feet were 70 per cent of the length of the upper leg bone, similar to chimps. Our feet are 55 per cent of the length of our upper leg bone. *H. floresiensis*' strange foot-to-leg proportions meant that it had to lift those feet higher and bend its knees further back than we do, just to get ground clearance. 'Clown-like' is how Bill Jungers has described it.

Even the bones in the feet are of an unexpected form. Those on the outside edge of the foot are long and curved. The forefoot, which comprises the metatarsals and toe bones, is disproportionally long compared to ours. The toes are curved and strong, and they do not have the 'hourglass' shape that ours have. All in all, these characteristics suggest to Bill Jungers and Professor Dan Lieberman (Harvard University) that the feet apparently lacked the spring-like mechanism that we use in running, and the curved toes may have posed another hindrance to running. They think that *H. floresiensis* must have evolved before the characteristics for running appeared in our genus.[26]

I've mentioned the relatively short legs of LB1. The effect of this was that their arms extended much lower than ours, so LB1 would've looked somewhat ape-like in this respect, though not as exaggerated. These body proportions are similar to the two-million-year-old *Homo habilis* from Africa and even the 3.7-million-year-old *Australopithecus afarensis* ('Lucy'; 'The First Family').[27]

H. floresiensis' shoulders, too, were rather odd, and different from ours. Professor Susan Larson (Stony Brook University) and colleagues studied the shoulder bones of *H. floresiensis* in comparison to modern humans, Neanderthals, *H. ergaster* and some three-million-year-old australopithecines. They found that, far from having a modern human shoulder, *H. floresiensis* was similar to *H. ergaster*. Both had shoulders that hunched forward and upwards, so the neck would look deep-set.[28] We soon learned, too, that the 1.8-million-year-old hominins excavated at Dmanisi, Georgia had this type of shoulder.[29]

This form of the shoulder turned up again in the recently discovered *Homo naledi*.[30] Elen Feuerriegel, at the time a PhD candidate at the ANU School of Archaeology and Anthropology, was one of the archaeologists excavating the *H. naledi* bones. She had just returned from one of her fieldwork trips when she heard me describing the *H. floresiensis* shoulder to a colleague. Thinking that the morphology sounded familiar, Elen showed me her digital scan of the *H. naledi* shoulder. We pored over this together and Elen showed me how similar *H. naledi*'s shoulder was to that of *H. floresiensis*.

One of the impacts of having a shoulder like *H. floresiensis*, *H. ergaster*, the Dmanisi hominins and *H. naledi* was that their arms could not rotate like ours. Their shoulders would have worked just fine for most day-to-day activities,[31] but they probably could not throw using an overarm action. This might seem a trivial point, but it is important because, along with the progression of endurance running, the development of high-speed throwing is thought to have been an important selective advantage in the evolution of modern humans.

With a capacity of 426 cubic centimetres, *H. floresiensis*' brain was grapefruit-sized. What, though, of the likely cognitive abilities? Absolute brain size does not directly measure the cerebral capabilities of an individual; rather, it is the complexity and organisation of the brain that is important. Brains, of course, are not preserved in ancient fossil remains, but marks on the inside of fossil skulls sometimes reveal the presence and the form of arteries and of convolutions of the brain. Incredibly, such marks show up on LB1's skull. Professor Dean Falk (Florida State University) is an expert at interpreting these brain convolutions. She and her colleagues studied these marks and found that the skull of LB1 housed a relatively large frontal lobe.[32] In *H. sapiens*, this part of the brain is associated with capabilities for planning, learning from mistakes, multitasking, and passing on knowledge from generation to generation. In some other respects, LB1's brain was like those of *H. erectus* and *A. africanus*.

When was *H. floresiensis* around? Until 2016, the layers in which *H. floresiensis* bones had been excavated had been dated to between 18 000 and 95 000 years ago. Intensive archaeological work in Liang Bua cave over the last few years, however, has shown that the charcoal used to date the bone-bearing layers back in 2003 was not what we call 'in association' with the bones after all. What is now recognised is that the *H. floresiensis* bones were in a natural pedestal up to 4 metres wide that was truncated from the surrounding area. The charcoal samples, although physically very close to the *H. floresiensis* bones, had nevertheless been excavated from the other side of this truncation, and so they do not represent the age of the layers in which *H. floresiensis* was discovered.[33]

The team has now dated three *H. floresiensis* lower-arm bones using the uranium-thorium method. When bones lie in sediment they absorb uranium from groundwater, while the thorium in the bone decays. These amounts can be measured, and from this researchers can obtain an age range for the bones. The age of the LB1 lower-arm bone is between 71 000 and 87 000 years old. Two other individuals' arm bones are slightly younger, at between 66 000 and 71 000 thousand years old, and 52 500 and 66 000 thousand years old, respectively. So the first known appearance of *H. floresiensis* on Flores was around 90 000 years ago, and their last appearance could be as recent as around 53 000 years ago. Stone tools from the Liang Bua excavation, however, go back 190 000 years,[34] indicating that hominins had been in the cave from at least that time.

What do these dates tell us about when *H. floresiensis* arrived on Flores and when it became extinct? Unfortunately, it's not possible to say. Just because their bones are not found in Liang Bua after 53 000 years ago does not mean *H. floresiensis* went extinct at this time. Their remains could be discovered anywhere on the island and these could be dated to earlier or later periods than *H. floresiensis* in Liang Bua cave. Besides, the exact point of extinction or arrival of any species can be hard to pinpoint. The chances of finding the very first *H. floresiensis* individual that arrived on the shores of Flores, or the final survivor, are vanishingly small. And how would we know if we'd found these? All that we can say about *H. floresiensis* is that they visited this cave on and off between 53 000 and 87 000 years ago. Or, if the stone tools discovered in the excavations were made by *H. floresiensis*, then we can extend the species' visiting period back to 190 000 years ago. Tantalising, though, is the more recent discovery of hominin remains 74 kilometres away in the So'a Basin, which I will talk about in chapter 4.

We only know of *H. floresiensis* from one cave on one island, but from this we can say a couple of things about its activities. One of these involves stegodonts. Very high concentrations of stegodont bones and teeth were found in the excavations. This needed some explaining, as stegodont remains do not usually show up in caves, so Mike Morwood called in Gert van den Bergh (University of Wollongong). Gert has extensive knowledge of prehistoric animal fossils and is a specialist on the stegodont bones of Flores. Carefully sorting, counting and analysing the bones and teeth from Liang Bua, Gert and his team discovered something odd. Rather than being a random assortment of adult and juvenile bones, as you'd expect, the bones and teeth were almost all of young animals. Of the forty-seven individuals, 94 per cent were juveniles, and of these, 23 per cent were newborns.[35] It seemed that *H. floresiensis* was either hunting these young wild animals or scavenging their carcasses.

But I wonder, could the tiny hominin really fell a stegodont, even a young one? The Liang Bua stegodonts were the size of a water buffalo—about 400 kilograms, Gert estimates. So even a juvenile would be a challenge to take down. Perhaps *H. floresiensis* not only hunted but also worked cooperatively in a team. But still, it's a fairly brave individual who gets between a newborn and its sizeable mother. In fact, Gert and colleagues suggest other possible explanations. The young ones might have succumbed to starvation through periodic droughts or food shortages, or become trapped in pools of standing water in the cave, so being easy pickings.[36]

The other activity attributed to *H. floresiensis* in the cave is making tools from stone. Dr Mark Moore (University of New England) and colleagues have long studied the stone tools from Liang Bua. Archaeologists who have specialised in stone tool technology can tell us a lot about how a tool is made, even the sequence in which flakes have been struck off the rock. What Mark and the team found was that large flakes were struck off specially selected stone from some source outside the cave, possibly from the nearby riverbed, then carried into the cave, where they were further shaped.[37] The types of stone tool in the cave do not change during the entire period of occupation. At a particular time, however, the *type* of stone changed. To Mark Moore and colleagues, this change indicated the arrival of modern humans at Liang Bua cave. Thomas Sutikna and colleagues have recently dated that occurrence at around 46 000 years ago. They, too, suggest that the change in the preferred stone type from this time onwards indicates that modern humans were using Liang Bua cave.[38]

But can we be sure that *H. floresiensis* made the tools that were excavated from the same levels as the *H. floresiensis* bones? My caution stems from a salient lesson learned during my undergraduate years. I was researching the circumstances of the discovery of hominin fossils at Olduvai Gorge in Tanzania. Here, in 1959, the palaeoanthropologist Mary Leakey discovered a hominin skull, *Paranthropus boisei* (see Appendix A). The skull was found on the surface among stone tools. *Paranthropus* was assumed, therefore, to have been the maker of the tools. But a fossil hominin with a somewhat larger brain size of 600 cubic centimetres—50–100 cubic centimetres greater than *P. boisei*—was discovered in the same area ten months later; some hand bones were also discovered here.[39] John Napier, a physician, primatologist and palaeoanthropologist (King's College London School of Medicine and Dentistry), studied the hand bones and found that the individual had a hand capable of a power grip, and possibly a precision grip too.[40] The power grip is when the fingers (and sometimes the palm) clamp down on an object, with the thumb creating counterpressure; examples of the power grip are gripping a hammer, opening a jar using both your palm and fingers, and performing pull-ups. The precision grip is when the fingertips and the thumb press against each other; examples include holding a nail when hammering it, writing with a pen, and hand-sewing. Napier was in no doubt that the new species could use naturally occurring objects to its advantage (power grip), but he was less certain about the capacity for toolmaking (precision grip). Louis Leakey and colleagues nevertheless assumed that the larger-brained of the hominins, with its particular hand structure, is the more likely of the two species to have had the ability to make tools, therefore naming the new species *H. habilis*.[41] *H. habilis* is sometimes nicknamed 'Handy Man'.

My take-home message from this history is that there were two hominins in the same area in Olduvai Gorge, each found in association with stone tools. Who made the tools? We cannot be sure. Either one of those two species made them, or perhaps both species were responsible, or neither of them. That is, we can make inferences, but we need to remain open-minded when declaring just who made tools that are found with hominin bones.

Can we infer, then, that *H. floresiensis* made the tools excavated at Liang Bua? Unlike most other *H. floresiensis* researchers, I think we should leave this as an open question. We simply cannot know the answer, even if we have no likely suspects at this time apart from *H. floresiensis*.

It must have been good to have had such a commodious cave in which to live, eat and prepare stone tools. However, Liang Bua nonetheless may have been a dangerous place—some of the excavated bones belong to komodo dragons. Still around today, this animal is a somewhat lumbering creature, but appearances can be misleading. It can run at incredible speed and attack with ferocity. Once it has spotted its prey, it is too late for that unfortunate animal. Tourist guides are careful to keep their groups at a safe distance from these animals, even from the sleeping ones. That said, we don't yet know if the giant lizard was a danger to *H. floresiensis* in Liang Bua cave, because it is not clear if the komodo bones found in the excavations were in the same layers as the *H. floresiensis* bones. It's possible the hominins and komodo dragons were in the cave in different eras—for *H. floresiensis*' sake, let's hope so.

There is something else to note about Liang Bua cave: a great photo opportunity awaits visitors. From an elevated balcony next to a large stalagmite above a boulder-strewn area at the back of the cave, you can snap a picture of the whole panorama, over the excavation sites towards the entrance. Most photos you see of the cave are from this vantage point, looking out to the entrance. But once you have taken the photo, take a moment to turn to your right to face the cave's rear wall, then scan down to the boulders at its base. Between two of these boulders is a deep, black, 1 metre-wide gap.[42] Few visitors know of this place, and few would realise that this is a naturally formed entrance to an underground cave. When Carol Leftner showed me this entrance during the Gadjah Mada palaeoanthropology conference field trip in 2007, my mind raced: What's down there? What archaeological mysteries lie within? Hominins might have slipped or tumbled down ... the chamber should be excavated! Well, I wasn't the first to think of this opportunity.

In 2006, an Indonesian–Australian team reconnoitred Flores for caves suitable for their palaeoclimate research project.[43] The group comprised Professor Wahyoe Hantoro and Bambang Suwargadi (Research Center for Geotechnology, Indonesian Institute of Sciences); Honorary Professor Mike Gagan, Dr Linda Ayliffe and Neil Anderson (Research School of Earth Sciences, Australian National University); Garry K Smith (Newcastle & Hunter Valley Speleological Society); and Associate Professor Russell Drysdale (Environmental and Climate Change Group, University of Newcastle). Wahyoe explains the team's objectives:

> An important goal for our team was to help the archaeologists find out why the Hobbit disappeared from Liang Bua. Speleothem records can help tell if they left because of a sudden change in climate, or perhaps because of a big volcanic eruption that damaged the landscape. The team was hoping that the lower chamber might provide good samples for this work. There are a few stalagmites down there, but they are not in good condition. Fortunately, the team is making good progress to answer this question using speleothems from Liang Luar cave, which is located only 1km away from Liang Bua.[44]

It was Garry K Smith who discovered the tunnel when the team visited Liang Bua. The team then sought local permission to descend. Mike Gagan reminisces:

> Garry took a rope, hitched it up and abseiled down a 60-degree slope that descended 23 metres. He found a whole chamber hidden immediately below Liang Bua [the chamber is now referred to as Liang Bawah, or 'Cave Underneath']. He took measurements—the chamber is 430 m^2—and went quiet for a while. Then he yelled out that he'd found a bone.[45]

Not from a longed-for hominin, though. On descending, Linda Ayliffe, a geochemist specialising in geochronology dating (and an experienced caver), found only the bones of rats, pigs and bats, although she also found seventeen stone tools on the cave floor. Over the course of another two days, the group's small excavations revealed more than 220 well-preserved bone elements belonging to giant rats, pigs, primates and bats. The bones were found buried shallowly in rubble at the base of the shaft and around a 5-metre-high mud mound in another part of the chamber.[46]

Back in Australia, Linda dated samples of the calcite taken from the surface of some of the bones. She discovered that they ranged from 240 000 to just 3000 years old,[47] revealing the potential for other ancient bones to be discovered in the chamber. This galvanised Mike Morwood and the Liang Bua team to descend into Liang Bawah in 2007 with the palaeoclimate

Top: The team that worked in Liang Bawah in 2006. Image by Garry K Smith (July/August 2006). Reproduced courtesy of Garry K Smith.

Bottom: Mike Gagan in Liang Bawah. Image by Garry K Smith (July/August 2006). Reproduced courtesy of Garry K Smith.

research group. Garry K Smith describes how the non-abseilers accessed the lower chamber:

> Those on the 2007 trip who could not abseil and prussik were taught on a rope tied off and overhanging the balcony. I think Jodie Rutledge did most of the teaching to abseil because Neil and I were mainly tied up with inserting the permanent anchor bolts for the abseil down to the lower chamber. Those anchor bolts were removed at the end of the trip as we did not want to encourage people to go into the lower chamber.[48]

Garry adds that

> one other aspect which could be a danger to people entering the lower chamber is the possibility of carbon dioxide building up to dangerous levels. It is a hazard in many caves which are deep and don't have through airflow. The lower chamber is such a scenario and there is a possibility that at certain times of year—depends on atmospheric conditions, temperature and barometric pressure—[it] may allow a build-up to life-threatening levels. I have had a lot of experience with foul air containing elevated carbon dioxide and reduced oxygen levels, so was very much aware of the signs if I encountered it. This meant that on the first entry, I was ready to prussik straight back out if foul air was encountered.[49]

Later that year, Thomas Sutikna excavated to a depth of 1.5 metres at the base of the talus slope, looking for something that could be dated, or any other interesting things. 'We found only rubble and recent items, including a complete pig skeleton. Another area was all mud and clay and difficult to excavate,' he reports.[50] A third excavation, elsewhere in the chamber, yielded just limestone rubble.

Emma St Pierre (University of Queensland), Kira Westaway and colleagues, a group comprising Indonesian and Australian researchers, undertook further work in this chamber. They dug a test pit of 1 square metre to a depth of 130 centimetres, finding rat and bat bones but no stone artefacts. There was no evidence that humans or hominins had ever used the chamber.[51] Together with the earlier work, their results suggest that things may have accumulated in this cave from being washed into the sinkhole, rather than accumulating from humans or hominins using the underground chamber.

Perhaps old hominin bones might one day be found among the debris, but it seems the chances are slim. Also, the lower cave is a potentially dangerous environment in which to work, as we have heard from Garry K Smith. Mike Gagan recalls that it was extremely hot down there, and Linda Ayliffe remarks that researchers can only work there for a few hours at a time. The descent into the chamber is also steep and narrow, with loose rock underfoot.

'That was the worst thing—you don't want rock falling on you,' says Linda. Absolutely not, but the equal worst thing for *me* would be the massive spiders Linda describes: 'Bigger than your hand, all over the cave walls as you descend, spiders all around you.'[52]

Better, perhaps, to concentrate on Liang Bua for the time being. And this is what is happening. Jatmiko tells me that work continues each season in Liang Bua cave. The project is a collaboration between ARKENAS, the University of New England, University of Wollongong, Smithsonian Institute (USA) and Lake Head University (Canada). The objective is to make Liang Bua cave a centre of study for archaeology, human evolution and environment, part of a strong vision for the future.

Thomas Sutikna and the team see the training of the next generation of archaeologists as key. Each season, a number of students from different institutions come to gain practical archaeological experience. Planning is in place: the unexcavated side of the cave will be preserved for the next generation of archaeologists. 'They will be better than us one day,' says Thomas.

Nicknaming *Homo floresiensis* 'the Hobbit'

Tiny little *H. floresiensis* with its outsized feet was quickly nicknamed 'the Hobbit', a reference to a fictional race that appears in JRR Tolkien's *The Hobbit* and *Lord of the Rings* trilogy. One lunchtime several years ago, in the tearoom of the School of Archaeology and Anthropology at ANU, I asked Colin Groves if he knew how this nickname came about. He said that the day *H. floresiensis* was announced, he and Tim Flannery (then Director of the South Australian Museum) were talking prior to being interviewed on the ABC. The first *Lord of the Rings* film had just been released, and one of the men said *H. floresiensis* was a dwarf. The other replied, 'It's not a dwarf in the Tolkien sense. It's a hobbit, isn't it?' They then used this term over the airwaves.

When I asked Colin who in that conversation said, 'It's a hobbit, isn't it?', he hesitated and said he didn't know. Now, anyone who knew Colin would remember his prodigious memory and self-effacing manner, and would surmise, as I do, that it was he who applied the term 'hobbit' to *H. floresiensis*. Otherwise, he would have quickly given credit to the instigator. There could, however, be other claimants to this piece of history, as Mike Morwood also used the term in an interview that same day.

So the little mystery remains: who coined the apt nickname for *H. floresiensis?*

Postscript

An anonymous reviewer of a draft of this book kindly provided further insight into this question. The reviewer related that, upon visiting one of the *H. floresiensis* team at the University of New England, they were shown an early draft of the article the team was then preparing for submission to *Nature*. In that manuscript, the species name was *Sundanthropus hobbitus*. This new information shows that the name 'hobbit' was already circulating among the discovery team at that time. It was clearly more than simply an appealing nickname.

2

Controversy from the start

'It's crystal clear—it's a modern human, if one with many problems,' claimed Dr Alan Thorne.

'It was totally outside the range of modern human variation. There was no way it could be modern. Very, very clear-cut,' said Professor Peter Brown.

Others exclaimed:

'It just invites tremendous scepticism.'

'I don't believe it.'

'There is only one skull and that's not proof of anything at all.'

'Entire careers are at stake.'

'One group is going to be entirely wrong.'[1]

After news broke of the *H. floresiensis* discovery, it rapidly became clear that this species would be very controversial.

This would not have been wholly unexpected. In their announcement of the new species, Peter Brown and colleagues, having carefully examined all the *H. floresiensis* bones, came up with two possible explanations for them.[2] One was that *H. floresiensis* was a dwarfed form of *H. erectus*. According to this scenario, it was assumed that *H. erectus* had somehow arrived on Flores from Java,[3] and that an insufficient amount of food or other conditions on the island led to an evolutionary response to dwarf. Peter Brown and colleagues' alternative idea was that *H. floresiensis* was a remnant population of some very early hominins, such as those that lived in what is now Africa a million or so years ago.

Science proceeds by researchers questioning and testing others' hypotheses and conclusions; this is the normal and appropriate procedure. And so researchers would have begun asking questions such as 'Did the team look at all possible explanations for the Liang Bua bones?' or 'Did they interpret their information correctly?' or 'Could the bones belong to *Australopithecus* or to some already known species of *Homo*?'

But then, suddenly, something came in from left field. On 31 October 2004, just three days after the announcement of *H. floresiensis*, an opinion piece appeared in Adelaide's *Sunday Mail* newspaper. Referring to the discovery, but with a curious lack of supporting images, Professor Maciej Henneberg (University of Adelaide) claimed that: 'Dimensions of the face, nose, and jaws were not significantly different from modern people ... but the measurements of the brain fell a long way below the normal range.' He added: 'The bell rang in my head ... the skull is of a microcephalic individual—there is not a single significant difference between *Homo floresiensis* and a microcephalic from ancient Crete.'[4]

Maciej Henneberg and colleagues were to expand on this, suggesting that LB1 had a brain size equal to the smallest of any human ancestor, while its teeth were similar to modern human teeth. They noted that LB1 had a normal-sized face but a brain size equal to early hominins from three to four million years ago, yet the species had the intelligence to make stone tools. Its body size was, 'considering its dating, way off the scale'.[5] They discounted the idea of island dwarfing for *H. floresiensis*, reasoning that had *H. erectus* voyaged to Flores once, it could have done so regularly, rather than become isolated and subsequently dwarfing. They also viewed Flores as not such a small island as to generate dwarfing as an evolutionary response to that environment. So island dwarfing of *H. erectus*, in their view, was unlikely to explain the existence of *H. floresiensis*.

Interestingly, Henneberg noted that many of the skeletal comparisons in Brown and colleagues' descriptions of *H. floresiensis* were made with the australopithecines rather than with *Homo*. This led him to wonder whether the discovery indicated the presence of a hitherto unknown form of

Australopithecus that had left Africa at least a million years ago and reached the Indonesian archipelago.[6] Nevertheless, he and his colleagues did not test this thought-provoking idea.

Seeking an alternative explanation for the Liang Bua bones, Henneberg and Dr Alan Thorne proposed that the individual known as LB1 was a modern human with the medical condition microcephaly (literally 'tiny brain')—in essence, they were saying that this was not a new species at all. It was a contentious suggestion by well-known researchers.

We didn't know it at the time, but the microcephalic modern human theory was the first of several arguments that would be put forward to oppose the acceptance of *H. floresiensis* as a new species. Over the next few years, we would read about the use of Laron syndrome, cretinism and Down syndrome to explain the characteristics of LB1, not to mention connections with small-statured modern human populations. The discovery of *H. floresiensis* had really started something.

While challenges to the proposed ancestry of *H. floresiensis* were anticipated, it was quite unexpected that the loudest assertions were about how *H. floresiensis* was not a new species at all but a modern human. Those making such assertions had to explain away the obvious differences between *H. floresiensis* and us, the solution being to claim that *H. floresiensis*—or at least the individual LB1—was a modern human with a pathological condition.

But back to 2004. With so much buzz about the new species, and with the annual conference of the Australasian Society for Human Biology (ASHB) coming up, Dr David Cameron (University of Sydney) and I thought there was an opportunity to engage with a wide audience on the differing ideas about *H. floresiensis*. The conference program had already been finalised but we were able to add an extra session to focus on this new little species. David and I would present our work, as well as co-chair the gathering, while Professor Henneberg and Dr Denise Donlon (University of Sydney) would also take part.

For our presentation, I compared data from the skull shapes of different early human species and a couple of microcephalic skulls of modern humans described in archaeological excavation reports. In a nutshell, the results showed that *H. floresiensis* clustered with the earliest species in our genus,

H. habilis. And there was something else: it was notably separate from any modern humans, including the microcephalic skulls. It seemed that *H. floresiensis* (LBI) was most similar to early *Homo* skulls. Essentially, I laid out an evidence-based position that *H. floresiensis* was not a modern human with microcephaly.

My conviction became even stronger when I saw the images presented by Maciej Henneberg in his talk. Realistically, practically anyone looking at the images of the Minoan microcephalic skull and that of LB1 could point out the differences fairly easily. Some of these differences are what distinguish *H. floresiensis* from modern humans, so for clarity I'll run through these.

In comparing the two skulls, we can see that LB1's forehead is low while the Minoan has the high, rounded forehead typical of modern humans. LB1 doesn't have a chin; instead, the jaw slopes backwards. The Minoan, on the other hand, certainly has a chin. In side view, the LB1 face projects, while the Minoan's is vertical, just like ours. Also note that LB1 has bone protruding around the upper rim of the eyes, and a column of bone extends from the eyeteeth up towards the edge of the nose—these two structures are not seen on the Minoan.

Frankly, sitting there at the ASHB conference, I was astounded. The two skulls were just so dissimilar; in fact, the only feature they had in common was a relatively small head in relation to the face. Yet Maciej's presentation clearly had support from some people in the audience. Things got pretty lively as people tried to come to grips with the opposing views on *H. floresiensis.*

At one point, someone who was clearly frustrated or straight-up annoyed by the new species, called out, 'When I was at school we all knew that in evolution one species evolved into another.' In other words, their view was that human evolution follows a 'straight line' of ancestor and descendent species, with only a single human species existing at any one time, and so *H. floresiensis* could not have existed at the same time as *H. sapiens*. David Cameron jumped in to say that it was now generally agreed that human evolution is more like a tree, where multiple human species 'branches' can coexist at any point in time. As I recall, this was met by the comment that if *Homo floresiensis* was a new species, then we would have to unlearn everything we know about human evolution.

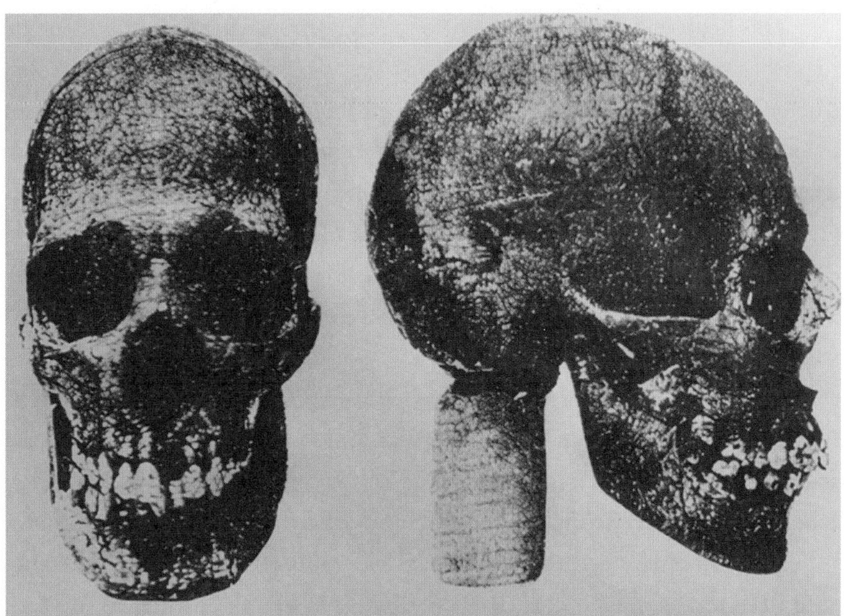

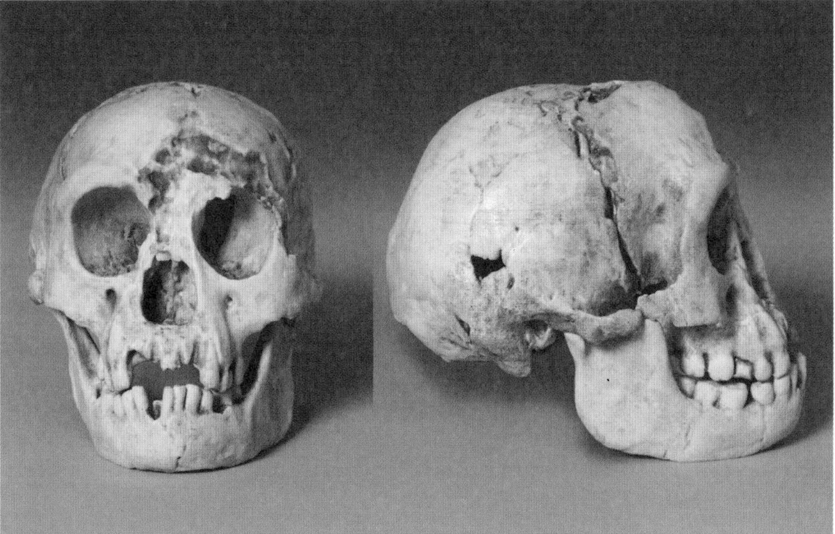

Top: Microcephalic skull excavated at Kato Zakros, Crete, and dated to about 2000 years ago. Image reproduced courtesy of Dr AN Poulianos.[7]

Bottom: The skull of *H. floresiensis* (LB1; replica). Images provided by Maggie Otto.

This heated exchange revealed to me that underlying the microcephaly hypothesis was a gut reaction against *H. floresiensis* as a new species. The acceptance of *H. floresiensis* was a problem if people viewed human evolution as a straight line, because *H. floresiensis* lived at the same time as modern humans. The concept of a linear evolution developed after Charles Darwin, who proposed in his book *The Descent of Man* that we are descended from an unknown form of Old World primate.[8] Ernst Haeckel, a German biologist, had come to the same conclusion. But there had to be some intermediary form between apes and us, and Haeckel even created a species name for this hypothetical being: *Pithecanthropus*. The 'straight line' idea of human evolution emerged as more primitive-looking hominin fossils were discovered and scientists began to think of these as evolving in a linear sequence from the most primitive hominins to modern humans.

Others saw some of these fossils as dead-ends on a more tree-like structure that nevertheless incorporated a dominant 'trunk' leading to modern humans.[9] That is, *A. afarensis* was followed by *A. africanus*, which was followed by *H. habilis*, which evolved into *H. erectus*, and we evolved from that species. There were two acknowledged divergences from this straight line: the robust *Paranthropus* coexisted in Africa with the australopithecines and early *Homo*; and the famous Neanderthals, *Homo neanderthalensis*, which lived in Europe and the Middle East from at least 200 000 to 30 000 years ago, existed at the same time as earlier populations of *H. sapiens*.

Since the 1980s, though, more and more discoveries have indicated that, in human evolution, different species persisted alongside others, while some species remained unchanged for long periods (see the species timeline that appears prior to chapter 1). Just as for other mammals in the fossil record—elephants, pigs, antelopes and so on—human evolution now looks much more like a tree of many branches than a straight line.

At the end of our ASHB conference session, I met Professor Richard Wright, who said that he was impressed with my skull analyses. Richard has a prodigious reputation regarding that sort of data analysis, alongside an impressive list of academic achievements, so when he raised the idea of working together on a project about the microcephalic–early hominin debate, I was most enthusiastic. Over dinner that evening, Richard, Professor

Colin Groves, Denise Donlon, David Cameron and I discussed the pros and cons of the idea. I had actually thought that any wider interest in this angle would fizzle out because it was so obvious that LB1 was not a modern human, but the others were much, much wiser than me. And Colin was already well ahead of the rest of us, brimming with ideas on how to get started. He asked Richard if he knew of any microcephalic skulls that we could examine. I assumed Colin would work on these, but to my surprise, he said, 'Oh no, I thought you could.'

And so, as a group, we started planning the work ahead. We would test the microcephalic hypothesis and work out to which species *H. floresiensis* was most closely related. This meant assessing the microcephalic question and the two ideas originally proposed by Peter Brown and colleagues: that *H. floresiensis* was a remnant population of archaic two-million-year-old hominins from Africa, or that it was a dwarfed *H. erectus*.

We had an ambitious project ahead of us, but we were starting from a good position. This was partly because we had already laid some of the groundwork, and partly because of the skills and experience of our team members. Denise had studied up on the *H. floresiensis* arm, leg, knee and foot bones, as well as the pelvis on which Peter Brown and colleagues had just published. David had worked out the most likely place *H. floresiensis* fitted on the human evolutionary tree. I had found that the microcephalic modern skull clustered with that of normal modern humans, while LB1's was most similar to early *Homo* skulls from Africa. And Colin was a powerhouse of knowledge on everything to do with evolution, taxonomy and hypothesis testing.

By early 2005, our team was hard at work. But within a few months, we had something else to add to the mix. Maciej Henneberg and Alan Thorne formally followed up Maciej's earlier newspaper article and ASHB presentation by publishing in the online journal *Before Farming*.[10] In their article, they argued that the LB1 skull and the Minoan microcephalic skull were both characterised by very small braincases, but their faces were 'within three standard deviations of the normal human range'. This meant that both faces were within the range of modern human face sizes, albeit at its outer edge. They cited small braincases and normal-sized faces as characteristics of microcephaly.

Henneberg and Thorne also disputed Peter Brown and colleagues' estimation of LB1's stature of 1 metre. Instead, they estimated that LB1 would have been 151–162 centimetres tall 'depending on method of reconstruction', although they did not detail how they worked this out. They pointed out that a modern human skeleton of similar stature had been excavated from another cave on Flores, Liang Toge, and dated to 3500 years ago.[11]

More theories about the origins of LB1 were to come, but before delving into those, let's take a closer look at the claims around microcephaly, particularly the impact this condition has on bones, which is the main evidence my team was dealing with. When I started reading up on microcephaly, I expected it to be a singular thing, with one set of characteristics that could be compared to the features of the *H. floresiensis* bones. But it wasn't that clear-cut. For example, I found that it's a very rare condition. It occurs in one in two million births in Scotland and one in 250 000 births in the Netherlands,[12] while in Sweden the incidence is predicted to be between one in 25 000 and one in 50 000.[13] Moreover, there are over 400 forms of microcephaly, not just one, as I had assumed. The condition is seen more commonly in regions with high rates of consanguinity—when close family members, such as cousins, have children—such as Turkey, Pakistan and some countries in the Middle East.[14] All microcephalics have a small brain that might be accompanied by other abnormalities such as short arms in relation to legs, short stature or joint defects. There may be cognitive impairment, where a person has trouble concentrating, remembering, learning new things, or making decisions that affect everyday life. The condition can range from mild to severe.[15]

The microcephalic skull on which Henneberg and Thorne had based their idea was excavated by Professor N Platon in 1962 from a grave in Malakari cave on eastern Crete. The find was dated to around 2000 BCE (that is, 4000 years ago), from the Minoan period.[16] Professor Poulianos, who reported the find, estimated that the skull belonged to a person of about twenty years of age.[17] The brain size was 350 cubic centimetres. Professor Poulianos concluded that the skull was that of a microcephalic. Fortunately for us, he had also published the measurements of the skull and I could now incorporate these into my analyses.[18]

I should also say that the discovery of the microcephalic skull on Crete was an exceptional find. It is rare to uncover bones in archaeological excavations that show individuals who had microcephaly. My research uncovered only five cases. Apart from the Minoan, there was a microcephalic from excavations in Sano cave, Japan, which I had used in my presentation at the ASHB conference. There was also one from ancient Egypt; one from a prehistoric burial ground in Peru; and one, a four-year-old child, from a cave on the Spanish island of Mallorca. Only the Minoan and Sano cave individuals were published with skull measurements that were necessary for our kind of analysis.

In 2006, Professor Teuku Jacob (Gadjah Mada University, Yogyakarta) and colleagues, including Maciej Henneberg and Alan Thorne, presented another explanation for *H. floresiensis*. They proposed that LB1, as well as other individuals found in the Liang Bua excavations, comprised an earlier population of small-statured modern humans. LB1, they claimed, showed signs of developmental abnormality, including microcephaly. They believed they had found 140 features on the LB1 skull—combined with those on two Liang Bua jaws—that were within the range of what we see in modern human skulls and jaws. In their view, some characteristics, such as the form of the eye socket and features at the back of the skull of LB1, are found in modern-day populations in South-East Asia, and in particular some people on Flores; here, they mentioned the Rampasasa villagers living near Liang Bua cave.

To test their idea, we needed to include the Rampasasa villagers in our study, but we had no luck in finding any information about them (note the postscript to this chapter). The next best thing we could do, then, was to use skull measurements of small-statured populations from elsewhere in the region. The closest to Flores were the people who live on the Andaman Islands in the Bay of Bengal, to the north-west of Sumatra.

In terms of what we had for comparison, we had amassed skull measurements of the Minoan and Sano microcephalic individuals, modern humans, a small-stature modern human population, the Liang Toge individual (mentioned in Professor Jacob's publication) and key early hominins. Now we had to find out how similar these were to LB1 in skull shape and thus obtain a better idea of the 'family tree' relationships between them.

Professor Richard Wright analysed the skull data. The results showed that the Minoan skull clustered with modern humans, albeit at the outer edge of the range; that is, its skull, although smaller than that of modern humans, was overall very similar to them. On the other hand, the shape of LB1's skull was quite different from the Minoan skull and all modern human skulls in our study. In fact, LB1 was close in shape to two 1.7–1.8-million-year-old hominin skulls from Africa: a *H. habilis* skull from Tanzania, and a skull from the Koobi Fora research area in Kenya whose species is still being debated.[19] That said, we were acutely aware that analysing skulls was not the end of the story. Other skeletal characteristics were vital when considering the origin of *H. floresiensis*, one of which was the relative length of the arm to the length of the leg (body proportions). Dr Denise Donlon therefore compared the arm-to-leg proportions of *A. afarensis*, *Australopithecus garhi*, *H. habilis*, *H. ergaster*, *H. floresiensis* and modern humans. She found that *H. floresiensis* limb proportions were similar to 2.5-million-year-old species *A. garhi* and dissimilar to any other species in the analysis.[20]

These outcomes led us to propose two possible scenarios for *H. floresiensis*. It could have evolved from a previously unknown early hominin that shares skull similarities with the Koobi Fora hominins and limb proportions with *A. garhi*, even though *H. floresiensis* is much younger than these species and was discovered in Indonesia, half a world away from Africa. Alternatively, it might have been in the process of evolving from *Australopithecus* to *Homo* when it wandered out of Africa.[21] These hypotheses were exciting because they potentially opened up a different way of thinking about what happened during human evolution, and they were also thought-provoking because they implied a movement of hominins out of Africa earlier than anyone had thought possible.

Meanwhile, Professor Dean Falk and colleagues were also testing the microcephalic idea for LB1 (as mentioned in chapter 1). Professor Falk is a specialist in the evolution of the brain, specifically how it led to the emergence of language, music, analytical thinking and even warfare; as part of her research, she has even studied Albert Einstein's brain![22] Surprisingly but happily, imprints of LB1's brain remained on the inside of the skull, and Falk and colleagues were therefore able to make a 3D model of the brain's form, with all its lumps and bumps—these replications are called endocasts. When she and her team compared the LB1 endocast with that from a microcephalic, they found that the two were quite different. Not only did they differ significantly in shape, they had other dissimilarities. For example, where the lumps of the brain just behind the forehead (called frontal lobes) were wide and flat on LB1, the microcephalic's frontal lobes were narrow and pointed. In fact, Dean and colleagues found nothing in LB1's endocast to suggest microcephaly.[23]

Neither the work of Dean and her colleagues nor ours dampened the microcephalic idea, however. Professor Robert Martin (Field Museum of Natural History, Chicago) and colleagues, experts in how human bodies grow and change in their proportions, did their own investigation into LB1.[24] They argued that if LB1was a dwarfed descendant of *H. erectus*, as initially proposed by Peter Brown and colleagues, then its brain would be much bigger than it was. Dismissing *H. erectus* as an ancestor of *H. floresiensis*, Martin and colleagues declared that LB1 had to be pathological. Developing this idea, they noted that LB1's short height, small and receding jaw, and some dental characteristics overlapped with those of microcephalics. The most likely explanation, in their view, was that LB1 could well be a microcephalic modern human.

Oddly, Martin and colleagues did not address one of the original and potentially key explanations for *H. floresiensis*: that it could be an archaic hominin species. They excused themselves for not testing this hypothesis with the stated reason that 'LB1 may derive from a more primitive (pre-*erectus*) population that cannot be addressed by consideration of modern human developmental abnormalities'.[25] I did not know what this meant. For more than a century, palaeoanthropologists have managed to work out the species of the fossils they discover. This process is not new. But such a position also supported my earlier thoughts about rocking the boat: that *H. floresiensis* as a new species was way beyond the comfort zone of some researchers.

It wasn't the last we heard of *H. floresiensis* being a modern human population with some sort of abnormality, or at least that LB1 was a modern human with an abnormality. Rather, 2006–07 turned out to be a pretty busy period regarding the *H. floresiensis* debate. New ideas and theories kept rolling in. Gary D Richards (University of California) suggested that the Liang Bua cave individuals were modern humans who developed a mutation in a particular gene, causing a growth hormone deficiency.[26] The population therefore became dwarfed in that island environment. Meanwhile, Professor Israel Hershkovitz (Tel Aviv University) and colleagues, noting the extremely short stature of *H. floresiensis* and its small brain, looked at whether these individuals were modern humans who had a condition called Laron syndrome.[27]

I had never heard of Laron syndrome. When I did an online search, I found that Dr Zvi Laron and colleagues had identified the syndrome in 1966. By 2015, it was estimated there were at least 500 people with Laron syndrome worldwide, although it is likely many cases remain undiagnosed.[28] The cases occur in the Middle East, mid and South Asia, and the Mediterranean, with the largest single group of affected individuals (about 100 people) living in southern Ecuador.[29] The genetic disorder occurs in some children of members of closely related families. It affects growth, resulting in short stature, among other complications. If the condition is not treated, adult males typically reach a maximum height of between 116 and 142 centimetres, and females grow to between 108 and 136 centimetres.[30] Keep in mind that LB1 is 106 centimetres tall, close to the smallest among Laron syndrome females.

I needed to know more and eventually tracked down a published photo of the head of a very young, unidentified child with Laron syndrome.[31] Taking a close look, I saw that the child had the high rounded forehead of a modern human infant, quite different from LB1's backwards-sloping forehead—the face did not project but was vertical like ours. This image of a child with Laron syndrome seemed very different from what I could see in LB1's skull. Emeritus Professor Zvi Laron describes this protruding forehead as a characteristic of Laron syndrome. Sufferers also have short arms and legs, small hands and feet, defective teeth that break easily and have many caries (holes), and a head that is disproportionally wide compared to the base of the skull.[32] *H. floresiensis*, however, has long arms but relatively short upper legs, very long feet, teeth which are in fairly good shape, and a skull that is widest towards its base. Laron syndrome was not shaping up well as a description of *H. floresiensis*.

Dean Falk and colleagues' interest was piqued too, and they searched the medical and clinical literature for further information. Having scoured the relevant publications, they listed the characteristics of people with Laron syndrome: short stature, protruding forehead, saddle nose, short face, broken and discoloured deciduous teeth with irregular growth and crowding of the permanent teeth, delicate arms and legs, small head circumference, and short limbs relative to trunk.[33] They, too, determined that the facial characteristics of Laron syndrome were quite at odds with those of LB1.[34]

Still, the 'pathology hypotheses' kept on coming. Professor Bill Jungers would wryly observe, 'Another day, another pathology.'

Cretinism was the condition that Peter Obendorf (Royal Melbourne Institute of Technology) and Emeritus Professor Charles Oxnard (University of Western Australia) proposed to explain *H. floresiensis*. Their view was that the Liang Bua group was part of a long-term modern human population that suffered from an iodine deficiency resulting in thyroid and growth problems.[35] They decided that the characteristic hips, arms, legs, shoulder bones and foot size of those suffering from cretinism were in the main similar to the features of the two *H. floresiensis* individuals, LB1 and LB6. Obendorf and Oxnard were very much aware that LB1's brain size was tiny (426 cubic centimetres), much smaller than that of any modern human. By suggesting that cretins of small-statured South-East Asian populations could have brain sizes of between 400 and 500 cubic centimetres, it seemed as if they were proposing that LB1 and LB6 were cretins from an unidentified, modern human, small-statured population.

Colin Groves was particularly curious about the claim that *H. floresiensis* individuals were cretins. While in Europe to undertake research, he took the opportunity to examine and measure ten cretin skulls and seven cretin skeletons at the Museum of Natural History in Basel. Along with Catherine Fitzgerald (ANU), he presented his findings at the 2009 ASHB annual conference, informing the audience that cretins' leg bones are far longer than those of *H. floresiensis*, that cretins' upper arms and feet are also quite different from those of *H. floresiensis*, that cretins do not have the small brain of LB1, and that cretins' teeth are in very poor condition (or missing) in comparison to the teeth of *H. floresiensis*. In total, Colin displayed nine PowerPoint slides documenting the skeletal differences between cretins and

H. floresiensis. He also found that Obendorf and Oxnard had misinterpreted some of the skull characteristics.[36]

The skull of a cretin differs to that of LB1 in having a high forehead, no mounds of bone above the eyes or the valley-like structure behind these, and the face does not project from under the eyes as LB1's does, although it projects just below the nose. Also, there is a chin, and the jaw is small in relation to the face.

Peter Brown roundly and comprehensively rejected cretinism as an explanation for LB1 and LB6. From the medical literature, he compiled the characteristics of cretinism across the whole body, forty-nine in all. Of these, thirty-three would have shown up on the bones we have for the two Liang Bua individuals, but not one of those cretinism characteristics was present. A quick look down his list said it all, really—the most prevalent word was 'No'.[37]

Karen Baab and colleagues then took up the mantle. I'd met Karen when we were working with other volunteers at the early hominin site of Dmanisi. Karen and colleagues used a computerised 3D skull-shape-analysis technique to test whether cretinism or Laron syndrome could explain the form of LB1's skull, or whether its skull was more similar to extinct species of *Homo*.[38] They found that LB1's skull shape was outside the range of variation in cretinism and was very different from those of humans with Laron syndrome. Instead, it was most similar to some fossil hominins.

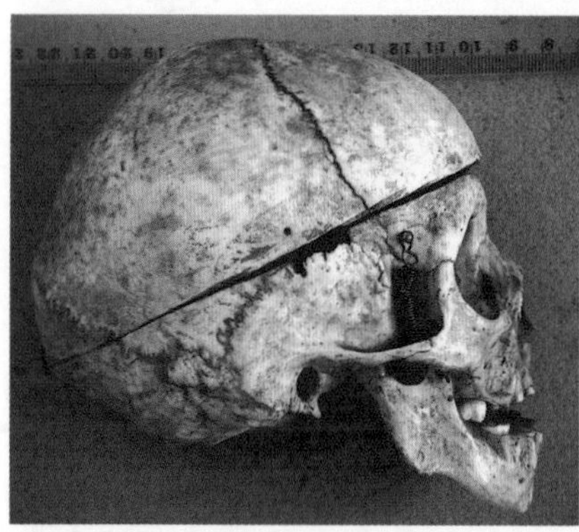

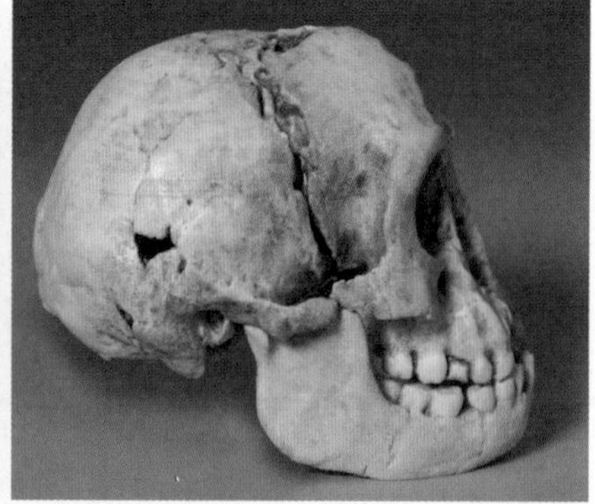

Left: Skull of a person with cretinism. The cretin's skull would have been cut at some time for demonstration or research purposes. Image provided by D Bulbeck and M Oxenham.[39]
Right: LB1 skull (replica). Image taken by Maggie Otto. (Photos not to scale.)

The issue of whether *H. floresiensis* was a modern human with pathology or a new hominin species continued to engage researchers' interest. It fuelled debate at international conferences in the United States, Indonesia and Europe, with multiple speakers presenting each view. At the same time, we were also getting to know more about the strange mix of features in *H. floresiensis*. As I related in chapter 1, at first we were confronted with a human-like creature that had some ape-like, australopithecine, and even early and later *Homo* characteristics. Now we were finding out that this little being had hunched shoulders and that its arms faced slightly forward compared to ours. Its feet were extremely long in proportion to its legs, and they were flat where ours are arched. There were also some odd curvatures in some of the toes. Ape-like wrist bones completed the picture.

Equipped with this clearer picture of what *H. floresiensis* would have looked like, and in light of our research, along with that of Dean Falk and Karen Baab and their colleagues, I entertained the idea that the 'pathology' papers would cease. But this was a short-lived thought. Microcephaly continued to be advocated as an explanation for *H. floresiensis*. RC Vannucci and colleagues, for example, performed some analyses on the skulls and endocasts of microcephalic individuals, comparing them to those of LB1. They concluded that LB1 was 'a brain-damaged individual, who suffered secondary microcephaly with associated severe motor disability (cerebral palsy) and likely mental deficiency'.[40]

Down syndrome was the next—and final—condition used in an attempt to explain LB1 as something other than a new hominin species. In 2014, Maciej Henneberg and colleagues proposed that LB1 was a small modern human with Down syndrome.[41] I decided to investigate this assertion, aiming to present my findings at the 2014 Australian Archaeological Association annual conference. I first looked up the skeletal characteristics of people with Down syndrome in the *Cambridge Encyclopedia of Paleopathology*. Then I listed the characteristics of LB1 and set about ticking any that also appeared in people with Down syndrome—nothing on that list got a tick. I also did the opposite, listing all the characteristics of people with Down syndrome and looking to tick those that also appeared in LB1, but again there were none.

Karen Baab and colleagues were also on the case. Examining the clinical literature on Down syndrome, they found that people with this condition have skulls often described as somewhat flattened at the back and base. The midface is hypoplastic, which means that the cheekbones and eye sockets have not grown as much as the rest of the face. The jaw is normal in size, but some individuals might have an unusually small chin. Down syndrome individuals also tend to have a short stature and a higher foot-to-thigh ratio than those who do not have the syndrome.[42]

Down syndrome skulls are of course modern in form, while LB1's is not. LB1's short stature of just over 1 metre and its disproportionally short lower limbs are primitive features also not seen in modern humans. All in all, Baab and colleagues found no support for LB1 being interpreted as a modern human with Down syndrome. They considered that LB1, and the other bones assigned to *H. floresiensis*, were better interpreted as a distinct species whose affinities lay with early *Homo*.

Dr Michael Westaway and colleagues also responded to the claim that *H. floresiensis* had Down syndrome.[43] They focused on jaw differences, and especially on one of the key things Peter Brown and colleagues had pointed out. It has long been established that early hominin species lacked chins. This is just one of several characteristics that show *H. floresiensis* is an archaic species, not a modern human.

I want to digress just for a moment on the subject of the chin, particularly as Michael Westaway and colleagues used the non-chinned jaws of *H. floresiensis* in their analysis. By 'chin', we are not just referring to the shape of the front of the jaw. Colin Groves alerted his students and conference audiences to an important feature of modern human jaws that had been described by Jeffrey Schwartz and Ian Tattersall.[44] This distinction is critical to the question of whether *H. floresiensis* is a modern human. The centre of the front of the modern human jaw is slightly raised, in the form of an upside-down 'T'; this bone reinforces the jaw. This specialised form of the jaw develops well before birth and is unaffected by the type of food that might be eaten or how vigorously we chew. But the Liang Bua jaws do not have this formation. We do not see this on *H. erectus* jaws or on other archaic hominin jaws.

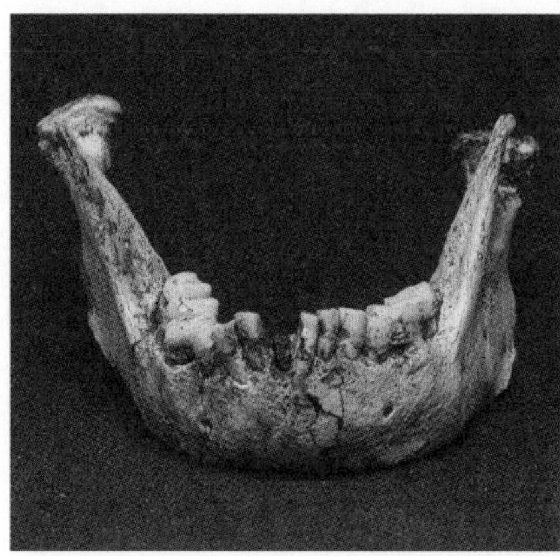

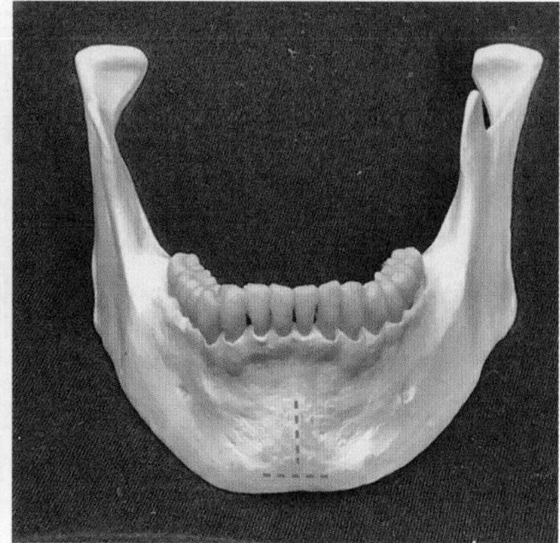

Left, the front of the *H. floresiensis* jaw, showing its smooth surface; and **right**, a modern human jaw (replica) showing the upside-down 'T' bone formation extending from under the incisors and across the base of the jaw. Images provided by Mike Morwood (LB1 jaw) and Debbie Argue (modern human jaw). (Photos not to scale.)

The reinforcement on the *H. floresiensis* jaws, however, is on the inside. Immediately below the front teeth is a ledge that slopes outwards and downwards. Below this is a bit of a depression before there is another ledge which stretches across the base of the jaw. Together, these structures reinforce the *H. floresiensis* jaw. Modern humans do not have these formations—the inside of our jaw descends straight down from behind our front teeth.

Based on the form of the *H. floresiensis* jaws, Michael Westaway and colleagues determined that neither LB1 nor LB6 could have been modern individuals, let alone have Down syndrome.

Even more evidence was to emerge that showed *H. floresiensis* was not a modern human with pathology. Antoine Balzeau (Muséum National d'Histoire, Paris) and Phillippe Charlier (Paris Descartes University) studied a computerised tomography (CT) scan of the skull of *H. floresiensis*.[45] As most people know, CT scans can show hidden features inside bones that can't otherwise be seen, even on close inspection. Scanning the skull of LB1, Balzeau and Charlier were able to see a variation in thickness across it, as well as the structure inside the bone. They also scanned thirteen microcephalic skulls from a number of museums, and some normal modern human skulls. Furthermore, from previous work they already had scans of twenty-six early hominin skulls, including *A. africanus*, *H. habilis*, *H. erectus* from Java, and

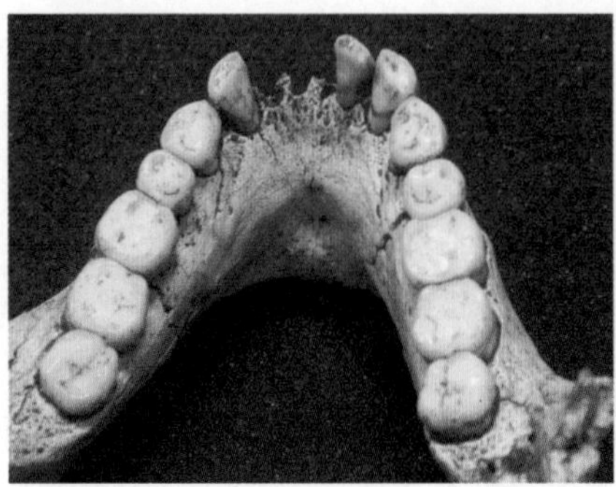
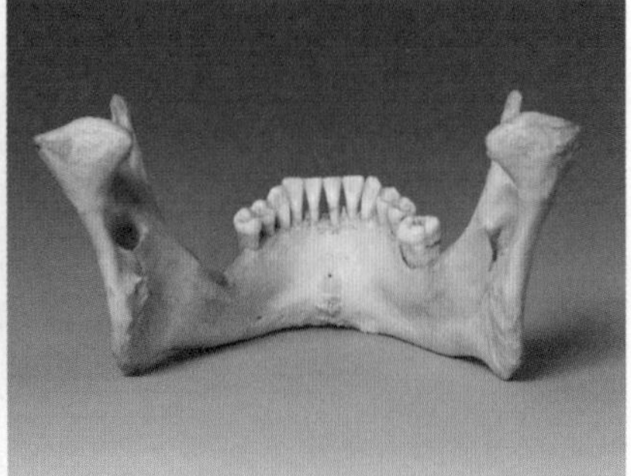

Left, inside view of a *H. floresiensis* jaw; and **right**, a modern human jaw (replica). The *H floresiensis* jaw is much smaller than modern human jaws. Images provided by Mike Morwood (LB1 jaw) and Debbie Argue (modern human jaw).

some Neanderthals. This was an impressive range of CT scans with which to test whether the LB1 skull was a microcephalic or if it fitted better with any of the skulls of the australopithecines or early *Homo*.

Balzeau and Charlier discovered much about LB1's skull structure. Most of this information was fairly complex, but the most relevant thing regarding our story was that all of the microcephalic skulls shared the internal skull bone conditions of modern humans, while LB1 did not. Balzeau and Charlier concluded that LB1 was not a modern human, with or without microcephaly. Disappointingly, however, they also found that the features they observed were not helpful for working out just which species *H. floresiensis* was closely related to.

So many hypotheses and so much controversy about this small being. Where does this leave us in the 'pathology' debate?

As I pointed out earlier in this chapter, it is normal procedure to test new scientific hypotheses. Most people would know that science works by examining all the facts concerning the question that is being investigated, and then determining which hypothesis best explains that information. Other researchers may then choose to test the hypothesis using additional data. Scientific progress is achieved in this way—nothing new here.

The pathology hypotheses generally focused on just the one set of *H. floresiensis* remains—LB1—with the implicit assumption that this was an abnormal individual in an otherwise normal population. But LB1 was not the odd person out here. The skeletal parts of the other individuals from the Liang Bua excavations represented the same (or even smaller) statured hominins as LB1. The (less complete) material we have of other individuals is all very similar to LB1.

Now, as we have seen, microcephaly and Laron syndrome are rare conditions. It would be reasonable to expect that, even if an archaeological excavation did reveal such a rare skeleton showing evidence of these syndromes, most of the other skeletal remains would be of a normal, non-pathological modern human population. But no bones of modern human stature or morphology have been recovered from the *H. floresiensis* levels in the excavations. Those who proposed these sorts of pathology for LB1 did not explain the *absence* of modern-looking, typical-statured human skeletal material in the deposit.

Finally, the proponents of the 'modern humans with pathology' explanation for the *H. floresiensis* population seemed not to realise something very significant about their proposals. At the time of this debate, the earliest evidence for modern humans in the region was 60 000 years ago, in Australia.[46] And yet, evidence for *H. floresiensis* went back at least 95 000 years. The hypotheses which implied that the entire *H. floresiensis* population consisted of modern humans, with or without pathology, unconsciously pushed back the date of modern humans' arrival in the region by 30 000 years. That would have been headline news indeed. But somehow, either the 'pathology' advocates did not realise this amazing implication of their work or they chose to ignore it, as it would have required a lot of explaining.

It is also worth pointing out that, since 2014, no new pathology hypotheses for *H. floresiensis* have been presented.

Time for some reflection. Clearly, the views on *H. floresiensis* were widely divided. This happens regularly in science, and it is expected that competing hypotheses will be tested. What was surprising was how strongly some researchers held onto the view that *H. floresiensis* was a modern human, even as the evidence for *H. floresiensis* being a new species became increasingly solid.

Professor Michael Lee (Flinders University; South Australia Museum), whom we meet again in chapter 3, shared the following with me on the polarisation of people's views in the *H. floresiensis* debate:

It can happen in science. How the debate about birds and dinosaurs played out is similar to what happened in the case for *H. floresiensis*. Before the 1990s there was strong but not overwhelming evidence that birds descended from dinosaurs. A small but vociferous bunch of people argued that birds did not descend from dinosaurs, that they derived from much more primitive crocodile-like reptiles that date back to 240 million years ago. But from the 1990s, all these feathered dinosaur fossils from China appeared, which absolutely, convincingly demonstrated that dinosaurs gave rise to birds. Anyone coming into this research field after the 1990s has no problem accepting that birds descended from dinosaurs. But some people made up their mind very early in the debate when the evidence wasn't conclusive. Once the evidence about the feathered fossil dinosaurs came in, these people still stuck to that initial opinion. They tried to reconcile the new evidence with their previous beliefs, rather than evaluating it objectively ... [and they did this] in one of two ways.

The first was that the newly discovered feathered fossils were true dinosaurs that had nothing to do with birds. These dinosaurs evolved feathers separate to birds evolving feathers. The other explanation touted was that these feathered dinosaurs were true birds and had nothing to do with dinosaurs. So they were birds that had evolved all these dinosaur traits. Neither explanation is very plausible. So I think what happens is that people in general, not just scientists, can form very strong opinions early on in the debate when the evidence is ambiguous, and when new evidence comes in they are very reluctant to shift their view. They try to mould and shape new evidence to fit their long-held beliefs, rather than the other way round ...

I think that's what happened with *H. floresiensis*. Before all the detailed anatomical evidence came to light, the idea that the bones were actually some sort of pathological modern human probably couldn't be completely discounted; a few people were attracted to that idea, making up their minds early on. When more and more evidence showed that *H. floresiensis* was a primitive species, those people had to try and shoehorn that evidence to fit the pathology idea and their explanations got more and more implausible ... and none of them fitted very well.

> Moulding the evidence to fit your ingrained ideas is something scientists should always be guarding against.[47]

Fortunately, while the pathology debates were in full swing, other researchers were testing the ideas presented by Brown and colleagues when they first announced *H. floresiensis*. We will now take a closer look at these ideas.

As mentioned earlier in this chapter, my colleagues and I were unable to find data for the Rampasasa villagers whom Professor Teuku Jacob and colleagues suggested were descended from *H. floresiensis*. This was a problem that beset a number of researchers who wanted to test this hypothesis. However, late in 2018, all was revealed—not through analyses of skull measurements, nor by looking at the bones of this group, but by that marvel of modern science known as deoxyribonucleic acid, or DNA. Serena Tucci (Princeton University) and colleagues sampled the DNA of thirty-two adult individuals from Rampasasa village, searching for gene sequences that did not match known DNA from modern humans.[48] Unknown DNA might have suggested that an unknown hominin could have contributed genes to the Rampasasa villagers' ancestors, but no such DNA was found. The Rampasasa small-statured group, Tucci and colleagues concluded, was not related to any unknown early hominin. Therefore, these villagers could not be descended from *H. floresiensis* (or any other hominin), and *H. floresiensis*, then, was not the ancestor of the Rampasasa village population, as proposed by Jacob and colleagues. This knowledge frees us to move on to the two supported hypotheses for the origin of *H. floresiensis* that are still being debated.

3

Fitting *Homo floresiensis* on the family tree

It's 2006 and I'm in a taxi, heading to Sydney airport to fly to Jakarta. I'm about to do something incredible: study the actual bones of *H. floresiensis.* This wonderful opportunity had come courtesy of Mike Morwood. He'd heard about the work that Colin Groves, Denise Donlon, Richard Wright and I produced that supported the theory of *H. floresiensis* as a new species, and had invited all of us to Jakarta. But due to other commitments, only I ended up flying out with Mike on a mission to meet *H. floresiensis*. I was beyond excited. This was a chance to understand more about just where the new species fitted into the family tree. I would be studying the bones in detail and working out which of the current theories about *H. floresiensis*' origin might fit the evidence.

I had never been to Indonesia before, so I was glad Mike would be flying with me—his inside knowledge and his contacts would make life much smoother. But first I needed to get on the plane. Worryingly, there was no Mike at the check-in. Or at the boarding gate. I suddenly realised that I was not really prepared for this, and panic started to rise. But then that familiar akubra appeared above a sea of passengers: Mike running in, just as the gate was closing, something about a passport. And just like that, we were on our way to meet *H. floresiensis*!

The *H. floresiensis* fossils were being kept at ARKENAS. Mike and I met with the Director, Dr Tony Djubiantono, and after a pleasant chat with him in his spacious office, we were taken into the large study room. Dominating the room was the safe where the bones were kept. Rokus Awe Due, who'd been the first to recognise that the small skull of *H. floresiensis* was an adult of an unknown species, opened the safe to reveal the boxes of bones. Then I did what any professional, sensible, cool and relaxed researcher does on such occasions: a jig! Mike laughed as Rokus, unaware of this little display, turned from the safe and reverently placed the boxes on a desk.

I remember seeing the skull and two jaws of *H. floresiensis* and almost not being able to breathe. My notes captured something of what I felt: 'I am in awe of the skull. I am so happy working on it ... It has such a striking mix of characteristics—Australopithecine, habilis, modern.'

Before I departed for Jakarta, Colin Groves and I had compiled a list of characteristics we wanted to know more about. Now, carefully examining the skull and jaws, I located these characteristics and made notes on them. Over subsequent days, I recorded other characteristics and drew sketches of these. Taking a whole raft of measurements completed that part of my study; all of this information would be invaluable for my PhD study.

My ultimate objective was to work out where *H. floresiensis* fits on the human evolutionary, or phylogenetic, tree. A tree in this sense is a branching diagram used to portray relationships among species. Its structure is derived from the fact that species shared common ancestors at various points in time, with its form based on shared similarities—in our case, the fossils of hominins. On any tree, the most closely related species share the largest number of unique traits: that is, characteristics not shared with others on the tree. Such groupings are interpreted as having shared an immediate common ancestor.[1] They are therefore more closely related to each other than they are to any other species in the analysis. In some ways, it is like anyone's family tree, except of course we are dealing with species rather than mums, dads, grandparents, second cousins twice removed ...

The process that sorts species into a phylogenetic tree is called cladistics. It's as challenging and formidable as it sounds, even though we use a software program to perform the analyses. I recall learning about cladistics in the second year of my human evolutionary course. I found it really complex and at the end of that year I thought, 'Well, I'll never have to think about that again.' Wrong. As I became more deeply involved in the study of human evolution, I began to see how useful the process was. Professor Colin Groves was an early adopter of the cladistics method and an expert in its use. Cladistics was a key technique I used in my PhD study.

Constructing the tree is a complex task. In the past it was laboriously done by hand and could take years to resolve. These days there are sophisticated programs that sort the species into relationships based on the data you provide. The tree is not considered 'the answer', though. Tests must be done to ensure the outcome is well supported. It is only then that a hypothesis of species relationships—a scientifically based idea or model—can be presented. Other researchers can test this model. And they usually do. That is how science progresses.

As I detailed in the previous chapter, we knew that *H. floresiensis* was not a modern human. So Colin Groves and I were particularly interested in examining the other main theories for *H. floresiensis*. One was that it evolved from a very early species in our genus,[2] like the ones that lived around 2–2.3 million years ago in Africa. This was potentially an explosive theory because it challenged the strongly held idea that it was the large-bodied, larger-brained *H. erectus*, rather than a smaller-brained earlier species, that spread from Africa to Java. This concept is so embedded as a model of human evolution that it has a name: Out of Africa 1.[3] The alternative idea was that *H. floresiensis* could have evolved from an unknown population of *H. erectus* that had arrived on Flores and become isolated from its mainland group. Without genetic exchange with that mainland group, *H. erectus* could have dwarfed into a new species. Peter Brown and colleagues had proposed these two ideas when they'd announced *H. floresiensis* back in 2004.[4]

As astonishing as the island-dwarfing idea seems, it did not come out of the blue. Dwarfing is a well-known evolutionary change that happens to some mammals that become isolated on islands. This is called the 'island rule'. Within a year of Peter Brown and colleagues' announcement of the new species, though, the *H. floresiensis* team had seriously rethought the dwarfing hypothesis, rejecting it when the discovery of more parts of the species showed just how archaic was *H. floresiensis*.[5] The primitive form of the *H. floresiensis* jaws also convinced Mike Morwood and colleagues that something was going on in the evolution of *H. floresiensis* that had nothing to do with *H. erectus*.

For record-keeping purposes and later reference, I took photos of the skull and jaws from all angles, and also of the particular characteristics of note. This was for Colin's and my use, so it did not matter to me that the images also included half my implements and my notes, or the background office furniture. Mike was thinking strategically, though. Picking up a black velvet cloth, he kindly taught me how to use this as a background to enable professional-looking photos, ones that he said would be good for publication. He had been dropping the words 'publication' and 'runs on the board' into our conversations for quite a while, and sure enough, at the end of my sojourn in Jakarta, Mike asked me if I would write a paper for an edition of the *Journal of Human Evolution* that would be dedicated to *H. floresiensis*.

Once I was back in Australia, and with Mike's offer to publish in the *Journal of Human Evolution* foremost in my mind, I got straight to work. I prepared my data for a cladistics analysis; loaded it into the software program; ran my analyses; and tested, then re-tested, the results. The resultant evolutionary tree showed *H. floresiensis* and the earliest species in our genus, *H. habilis* and *Homo rudolfensis*, on the same branch. In a joint publication, Mike, Thomas Sutikna, Jatmiko, Wahyu Saptomo and myself proposed that *H. floresiensis* descended from an early but unknown species of *Homo*. *H. erectus* was on a different part of the tree: its closest relative was *H. ergaster*. We tested to see if *H. floresiensis* could possibly fit on the branch with *H. erectus*, but it could not. We therefore rejected the hypothesis that the small, enigmatic bones resulted from island dwarfing of *H. erectus*.[6]

After we'd published our results, I received an email from Professor Bill Jungers, who is an expert on the postcranial bones (those below the neck) of chimps and other non-human primates, the australopithecines, early *Homo*, modern humans and *H. floresiensis*. More and more information was emerging about the postcranial bones of *H. floresiensis*, so Bill suggested he, Colin and I do a joint analysis that covered the characteristics of the arms,

Professor Colin Groves and me studying *H. floresiensis* at ARKENAS, Jakarta (2011). Image provided by Matt Tocheri.

legs and shoulders of various hominids. As it happened, Colin and I were already discussing this idea, so we jumped at Bill's suggestion.

In August 2011, having acquired generous funding from the ARC, and after nearly a year of planning and organising research permits, Colin and I headed off to visit the institutions that held the fossils we wanted to study. We travelled to South Africa, Kenya, Ethiopia, Tanzania, the Netherlands, Germany and Indonesia; Bill joined us in Jakarta to study *H. floresiensis* postcranial bones and provide insights into the postcranial characteristics of other species. Along the way, we collected extensive data on the characteristics of each fossil we examined, no matter how fragmentary—we measured and photographed everything.

To see and handle real, actual fossils, well, there's nothing like it for seeing nuances of form. Surprises can happen at any time. When we were studying the fossils of *A. afarensis* at the National Museum of Ethiopia, for example, we saw that the skulls had a fissure running along the length of the ear tube. Sure, it's not the most dazzling characteristic, but we had been looking specifically for this fissure on all the fossil skulls we studied because *H. floresiensis* had it. When it showed up on the *A. afarensis* skulls, we knew we had discovered something new: yet another *A. afarensis* feature that *H. floresiensis* possessed.[7]

By late 2013, after a few false starts, we had submitted our paper about the place of *H. floresiensis* on the human evolutionary tree to the *Journal of Human Evolution*. It had taken around eighteen months to perform all the analyses and write up the results. The journal's editors liked the way the paper was written, even asking us to do more work.

Enter Mike Lee. During 2013 I had accepted an invitation from Professor Gavin Prideaux (Flinders University) to give a talk on *H. floresiensis* at the Biennial Conference on Australasian Vertebrate Evolution, Palaeontology and Systematics (CAVEPS) in Adelaide in October 2013. I saw that there was something extra on offer at this conference. Being only too keen to hone my analytical skills, I signed up to a pre-conference workshop on how to use the Phylogenetic Analysis Using Parsimony (PAUP) software program for reconstructing evolutionary trees—Colin and I had been using PAUP to perform our analyses. Professor Mike Lee, whom I introduced in the previous chapter, and his workshop co-host, Associate Professor Matthew Phillips (Queensland University of Technology), specialise in these complex computer programs that work out where any given species sits on the evolutionary tree of life. Mike Lee is also an expert on reptiles and their evolution. Partway through the workshop, I had a revelation: the newer version of PAUP had far more potential than the one Colin and I had been using, so why not invite Mike to join our team. Mike ultimately took the lead in performing our analyses, and he was a co-author of the team's paper on where *H. floresiensis* lies on the human tree.[8]

When Mike analysed the fruits of the fossil-studying trips that Colin and I had taken, two equally supported trees emerged. One of them showed *H. floresiensis* and *H. habilis* together on a branch separate from the other species in the analysis. The technical term for this arrangement is a clade. The species that form clades are often referred to as sister species, which are interpreted as sharing a common ancestor that is not shared with any other species in the analysis. This led us to propose that *H. floresiensis* and *H. habilis* shared a unique common ancestor, making the *H. floresiensis* lineage at least as old as the *H. habilis* lineage—*H. habilis* is known from 2.35 to 1.65 million years ago.

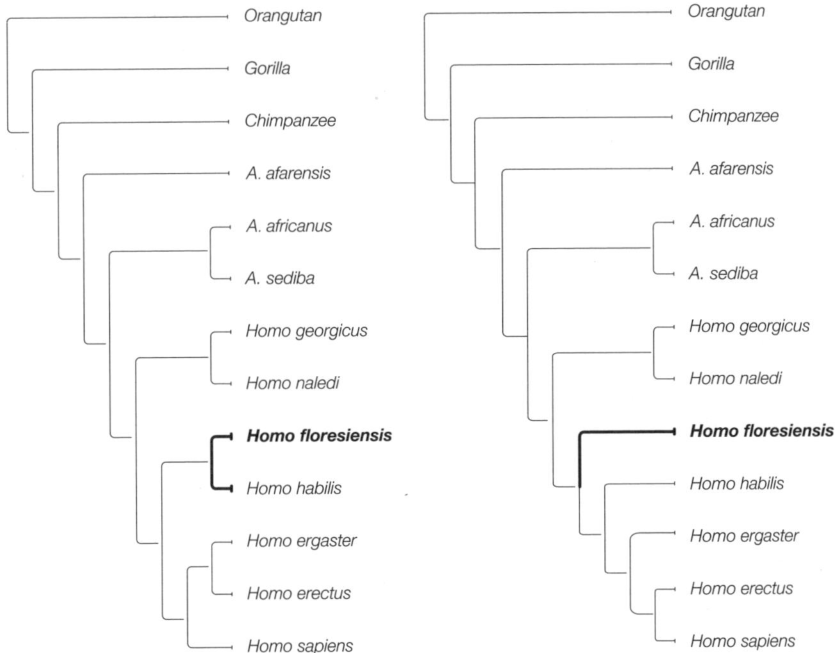

Left: *H. floresiensis* and *H. habilis* form a separate branch on the tree.
Right: *H. floresiensis* at the base of the group comprising *H. habilis*, *H. erectus*, *H. ergaster* and *H. sapiens*.

The other tree showed a slightly different arrangement of species, with *H. floresiensis* on a separate branch of the tree. Its position suggested to us that the *H. floresiensis* lineage could be even older than the *H. habilis* lineage. Both the phylogenetic positions implied an Early Pleistocene appearance for the *H. floresiensis* lineage.

In both cases, *H. floresiensis* was quite separate from *H. erectus*. In no case did they form sister species, as would be expected had *H. floresiensis* evolved from *H. erectus*. In one analysis, *H. erectus* was a sister species to *H. ergaster*; in the other tree, it was a sister species to *H. ergaster* and *H. sapiens*. We did a number of tests for the scenario that *H. floresiensis* evolved from *H. erectus*, but they all verified our trees and showed us that *H. erectus* was unlikely to have given rise to *H. floresiensis*.

These results strongly suggested that a very early but unknown species of *Homo*, or *H. floresiensis* itself (unknown in Africa as yet), left Africa and that its descendants made it all the way to Flores. This in turn supported one of the original ideas presented by Peter Brown and colleagues.[9]

It's an extraordinary idea: that something more archaic than *H. erectus* got out of Africa and made its way to Flores, where it held on for more than a million years after its closest relatives died out. It certainly challenges the Out of Africa 1 model of human evolution.

Ours wasn't the only work going on in this space. Another team used a technique called Bayesian statistics. In essence, the Bayesian approach is a measure of probability of the strength of evidence in favour of one model over another.[10] Using an extensive suite of characteristics of the skulls, jaws and teeth of twenty hominin species, the team led by Mana Dembo (Simon Fraser University, British Columbia) found that their best-supported tree was that in which *H. floresiensis* was on a branch leading to *H. habilis* and *H. rudolfensis*. This suggested that *H. floresiensis* was a descendant of pre-*H. erectus* small-bodied hominins, just as our team had found. The finding was pretty encouraging.

But then another team obtained a different result. Valéry Zeitoun (Université Paris-6, Sorbonne universités) and colleagues, using data from the calvarium—the part of the skull that encloses the brain—found that LB1 fitted on a branch of the evolutionary tree with *H. erectus* and *H. ergaster*.[11] The team were quick to acknowledge that support for this outcome was weak.[12] Nevertheless, they proposed that LB1 (the partial skeleton of *H. floresiensis*) was *H. erectus*, rather than a new species. How it became so small compared to other *H. erectus* individuals, they added, remained unanswered.[13]

Zeitoun and colleagues did not include information about the face, jaws, teeth or other body parts in their analyses, claiming that the australopithecine and early *Homo* characteristics in *H. floresiensis*' jaw, hand, foot and shoulders merely showed that the partial skeleton was not a modern human.[14] Other researchers, however, regarded the australopithecine and early *Homo* characteristics as very important evidence to consider when assessing the place of *H. floresiensis* on the human evolutionary tree.

So where did all that study leave us? Two cladistics analyses found that *H. floresiensis* likely stemmed from a very early hominin in our genus

Homo, and a third proposed that the *H. floresiensis* individual was a small *H. erectus* and rejected *H. floresiensis* as a new species.

Beyond cladistics, we could get a sense of what *H. floresiensis* might be by comparing its skull shape with those of other hominin species. Essentially, we want to see if the *H. floresiensis* skull is similar to the skulls of any other species. We can then make a call as to which species it might be most closely related.

One of the ways to objectively assess skull shape similarities and differences is through measurements. About half of Colin's and my time when studying *A. afarensis*, *A. africanus* and skulls of *Homo*, was spent taking copious measurements for just this purpose. You measure each skull of interest from specific landmark to specific landmark—landmarks, in this sense, are places on the skull (or jaw, tooth, leg bones and so on) that are anatomically recognisable places. The length of the skull, for example, is measured from a landmark between the eyebrows, called the glabella, to the area at the back of the skull that most protrudes. There are biological landmarks for many parts of the skull (the top, bottom, sides), for the face (eye orbits, nose opening, ear hole), you name it. Curvatures on the skull can also be captured. Once you've taken these measurements, you (patiently) load this data into a software program that will perform the required analyses. You can then find out which skulls are similar in shape and which ones differ significantly.

This is the kind of analysis I did for the ASHB conference I talked about in chapter 2, the one at which *H. floresiensis* was so hotly debated. Our team went on to do a comprehensive comparison of *H. floresiensis* with the fossils of *Homo* (including modern human microcephalics), *Australopithecus* and *Paranthropus*. I reported the results in the previous chapter, so I won't repeat the details here, except to say that we found the *H. floresiensis* skull was more like *H. habilis* and a 1.8-million-year-old skull from the Koobi Fora research area in Kenya, called KNM-ER 3733, than it was to *H. erectus*.[15] It is uncertain which species KNM-ER 3733 belongs to, although some researchers include it in *H. ergaster*.

A number of other research teams have also compared the shape of the *H. floresiensis* skull with other hominin skulls. Using a variety of analyses, Karen Baab and colleagues established that the shape of *H. floresiensis*' skull was what would be expected for a very small specimen of early *Homo*.[16] They found, as we did, that in some analyses *H. floresiensis*' skull clustered with KNM-ER 3733. Adam Gordon (George Washington University, Washington, DC) and colleagues found that *H. floresiensis* was most similar to one of the skulls of *H. georgicus*,[17] to KNM-ER 3733, and to a lesser extent to two *H. habilis* skulls. Because there was little similarity between *H. floresiensis* and *H. erectus* (Sangiran 17 from Java), Gordon and colleagues proposed that the ancestry of *H. floresiensis* did not include *H. erectus*.[18]

In contrast to these outcomes, a study by George Lyras (National and Kapodistrian University of Athens) and colleagues found that the skull shape of *H. floresiensis* was most similar to *H. erectus* and therefore suggested a close relationship between the two. Lyras and colleagues recognised that their results were different from the other skull shape studies, asserting that the disparity was due to the other teams using a dissimilar analytical technique. Unfortunately, they were unable to include some of the skulls of species that the other teams had found were closest in shape to the *H. floresiensis* skull, explaining that, in the case of *H. georgicus*,[19] the Dmanisi material was unavailable to them.[20]

The results of the skull shape studies, then, were somewhat inconclusive. Three were in agreement that the *H. floresiensis* skull was not shaped like *H. erectus*. They each concluded that *H. floresiensis* was unlikely to have descended from that species, instead suggesting that we were looking at an early hominin lineage for *H. floresiensis*. Another study, though, suggested that *H. floresiensis* probably evolved from *H. erectus*.

Peter Brown and colleagues gave us good initial descriptions and measurement data regarding *H. floresiensis*' skull to work with,[21] but we were eagerly awaiting a full description of the skull so that we could learn about all of its characteristics. So it was a most welcome addition to our knowledge when, in 2011, Professor Yousuke Kaifu (National Museum of Nature and Science, Tokyo) and colleagues published a detailed description of *H. floresiensis*' skull and face.[22] Kaifu, as his friends and colleagues call him, is a human evolutionist, and one of *the* experts on *H. erectus*.

Now armed with so much more information than was available previously, Kaifu and colleagues went on to explore whether *H. floresiensis*' skull and face could shed light on its ancestry. Assembling data from a comprehensive suite of fifty-four fossils that included *H. habilis*, *H. ergaster*, some later African hominins, *H. georgicus*, and *H. erectus* from Java and China, they tested whether *H. floresiensis* might have originated from *H. habilis*, *H. georgicus* or *H. erectus*.

Kaifu and colleagues identified twenty-one characteristics that are not seen in *H. habilis* or *H. georgicus* but occur in the *H. ergaster/erectus* group of skulls. They argued that, as these characteristics were absent from the earlier hominins and present in the later species, *H. floresiensis* must have inherited them from the later species. This would rule out the earlier species of hominins as ancestors of *H. floresiensis*. The small size of *H. floresiensis* could be explained, they said, by an unknown *H. erectus* or related form that arrived on Flores and dwarfed substantially in body and brain size in an isolated island setting. They left this question open to debate, however, noting that evidence from the skull alone could not solve the question of *H. floresiensis*.[23] Turning to the *H. floresiensis* teeth, Kaifu and colleagues examined twenty-six characteristics, many of which they noted evolved after *H. habilis* had evolved. This, too, would argue against an early hominin such as *H. habilis* being ancestral to *H. floresiensis*.[24]

For more clues, let's consider the jaws. When *H. floresiensis* was announced, we heard about some primitive traits in its jaws,[25] as primitive as we see in the 3.7-million-year-old *A. afarensis* jaw LH 4.[26] More details emerged when Peter Brown and Tomoko Maeda carried out a full study of the Liang Bua jaws.[27] The two researchers found that these jaws shared a distinctive set of traits that placed them outside the range of variation seen in *H. erectus*. The front and sides of the jaws, the buttressing structure inside the jaw (see chapter 2), and the form of the bony part of the jaw that links it to the skull, were similar to *Australopithecus* and early *Homo*. All up, when they considered the jaws along with the form of *H. floresiensis*' skull, its limb proportions, and its archaic wrists and shoulders, Brown and Maeda saw that *H. floresiensis* was in many respects closer to African early *Homo* or *Australopithecus* than to later *Homo*, including *H. erectus*.

This evidence suggested to them that the ancestors of *H. floresiensis* left Africa before the evolution of *H. erectus* as defined by *H. georgicus*. Brown and

Maeda believed on the basis of their analyses that *H. floresiensis*, a distinctive, tool-making, small-brained, australopithecine-like, upright-walking species, arrived on Flores *before* the arrival of *H. erectus* and *H. sapiens* in the region.[28]

Brown and Maeda went further, however, predicting that *H. floresiensis* might possibly have left Africa even before the evolution of the genus *Homo*. This is a radical idea, very similar to one of the scenarios my colleagues and I presented when we suggested that *H. floresiensis* could have been in the process of evolving from *Australopithecus* to *Homo* when it made its way beyond Africa (see chapter 2). Brown and Maeda and our own team, then, independently came to a similar conclusion: that it is entirely possible that a small, archaic hominin left Africa before two million years ago,[29] or 1.8 million years ago.[30]

Moving beyond the jaw, teeth contain a powerhouse of information. Readers will be familiar with how teeth and dental records are used in forensics. They are also one of the most informative components we can use in studying what happened in human evolution.[31] Even better, they are very tough and can outlast other bones in the fossil record, so we have many of them.

Brown and Maeda showed that *H. floresiensis* had a mix of archaic and modern characteristics in its teeth, but one thing in particular stood out. The premolars had a crown shape only seen in apes that lived way back in the Miocene (from twenty-three million to five million years ago),[32] and in the early australopithecines. By the time the later australopithecines and early *Homo* had evolved, this premolar shape had disappeared. The *H. floresiensis* premolars, then, have a very primitive form indeed, more primitive than *H. erectus*.[33] Compare this to Kaifu and colleagues' conclusions and we have two diametrically opposed views of the teeth of *H. floresiensis*.

We now move from the outside of the skull to the inside. Professor Dean Falk and colleagues looked at the issue of *H. floresiensis*' ancestry from the perspective of its brain. As we have seen, an imprint of the brain may be preserved on the inside of some hominin skulls. Falk and colleagues produced an endocast of *H. floresiensis*' brain by using 3D CT scanning. They then compared this to endocasts from the *H. erectus* skull from Java, four *H. erectus* skulls from China,[34] and endocasts from an australopithecine, a *Paranthropus* and some modern humans and chimps. They established that *H. floresiensis*' brain could not have been a miniaturised *H. erectus* brain, yet there were some similarities. So Falk and colleagues considered other physical aspects of *H. floresiensis*: its australopithecine-like pelvis, brain-to-body-size ratio, and characteristics of its upper leg bone, all of which are not expected in a miniaturised descendant of the larger bodied *H. erectus*.[35] They thought it possible that *H. floresiensis* might have been an island dwarf form, but equally that both *H. erectus* and *H. floresiensis* shared a common ancestor that was an unknown, small-bodied and small-brained hominin.[36]

It's now time to raise a principle that is basic to all science. The parsimony principle tells us to choose the simplest scientific explanation that fits the available evidence. That means regarding *H. floresiensis* that, all other things being equal, the best hypothesis is the one that requires the fewest evolutionary changes.

In two of the three cladistic analyses described above, *H. floresiensis* is on a branch with *H. habilis*, or with *H. habilis* and *H. rudolfensis*. Tests on those trees show that many more evolutionary changes would have had to take place had *H. erectus* evolved into *H. floresiensis*. As well, most of the analyses that compare skull shapes show that the skull of *H. floresiensis* is similar to that of *H. habilis* and two 1.5–1.8-million-year-old skulls from Africa, and quite dissimilar to *H. erectus*.

If *H. erectus* evolved into *H. floresiensis*, we have to accept that in doing so, a suite of archaic characteristics appeared, the likes of which had not been seen since the australopithecine and early *Homo* days. We call such characteristics 'evolutionary reversals'; that is, a characteristic reverts to a more primitive form over time. I'll explain what I mean by 'evolutionary reversals' by exploring some that are relevant to the *H. floresiensis*/*erectus* question.

Clearly, a massive shrinking of body and brain would have to take place.[37] Changes in specific aspects of the jaw and premolars, and, I contend, the feet and pelvis, are other examples. The *H. floresiensis* jaws are australopithecine-like in structure, something not seen in *H. erectus*, *H. ergaster* or *H. georgicus* jaws, all of which are heading towards the modern form. Why such a reversal would be selected for as *H. erectus* (putatively) evolved into *H. floresiensis*, even under the conditions found on an island, I do not know. Likewise, *H. floresiensis*' lower premolars are exceedingly and consistently primitive, similar to those of the australopithecines and early *Homo*.[38]

H. floresiensis has ape-like foot proportions, similar to the australopithecines: they lack arches on their feet, and they have a long forefoot and some long toes as well as an odd-shaped ankle bone.[39] We do not have foot bones for *H. erectus* or *H. ergaster* but we have something else: footprints. Around 1.5 million years ago, some hominins walked across a silty, sandy area near Ileret in northern Kenya. The footprints are like our own: we can see that the big toe is aligned with the foot, the heel is rounded and the sole is arched. These footprints are our earliest evidence for essentially modern feet in the human fossil record. They show that by 1.5 million years ago, some hominins had evolved an essentially modern foot function and style of walking. The size of the footprints is consistent with individuals who had the body mass and stature of *H. ergaster*, and it is presumed that it is this species that was walking along the lake's edge all those eons ago.[40]

Had *H. floresiensis* evolved from *H. erectus*, we would have to assume that *H. erectus* had the ape-like foot characteristics we see in *H. floresiensis*, or that, for some reason, the foot structure reverted to ape-like as *H. erectus* evolved into *H. floresiensis*. But we have never seen such an evolutionary reversal in human evolution. Why this would happen is puzzling when re-evolving long toes and a flat foot would impair walking performance.[41] Also, as we saw in chapter 1, the form of *H. floresiensis*' feet probably evolved before the characteristics for running appeared in our genus.

H. floresiensis' pelvis is australopithecine-like. We do not have a pelvis for *H. erectus*. Again, had *H. floresiensis* evolved from *H. erectus*, we either assume that *H. erectus* also had an australopithecine-like pelvis, or, if we do not think this was the case, we would have to argue that the *H. erectus* pelvis reverted to australopithecine-like as it evolved into *H. floresiensis*.

The brain of LB1 is only 426 cubic centimetres. This is closer to the brain size of apes, the australopithecines and some *Paranthropus* species. *H. floresiensis*' brain is smaller than any other species of *Homo*. It is even smaller than *H. habilis*, whose brain size ranges from 509 to 638 cubic centimetres. *H. erectus* brain sizes range between 813 and 1059 cubic centimetres (see

Appendix A). An evolutionary reversal in *H. erectus* to LB1's brain size would require us to accept that, in evolving to *H. floresiensis*, the *H. erectus* brain would have shrunk to half its size or even less than that.

It all comes down to whether we are comfortable accepting so many evolutionary reversals in one species when we have not seen this degree of reversal in the evolution of any other hominin species. As Bill Jungers has pointed out: 'Some modern humans (pygmies) have reduced greatly in body size repeatedly and independently throughout the world, without any evidence of evolutionary reversals to such primitive morphologies and body mass.'[42]

Of course, while evolutionary reversals are not known to have occurred to such a significant extent during human evolution, that doesn't mean they did not happen. They have certainly been a factor in the evolution of other species. When I asked Professor Mike Lee, who is an expert on the evolution of reptiles, especially snakes and lizards, if reversals have occurred in that domain, he replied:

> Yes, at least in reptiles it happens a lot. For example there are many lizard species that have completely lost their legs during evolution. Why do we assume that those lizards have lost their legs, rather than primitively never had them? The reason is that there is a whole lot of anatomical and genetic evidence that places these legless reptiles deep within the lizard branch of the evolutionary tree—they are true lizards. So we know that they are descended from lizards with legs. For instance, many of the legless lizards in Australia are geckos that have lost their legs. They chirp like geckos, they always lay two eggs just like geckos, their ears are just like geckos.
>
> We see many primitive traits of *H. floresiensis*, such as the jaw shape, short upper legs and so forth. Can we interpret these as reversals from a *H. erectus* condition? Well, to do that, we'd first need to find even more traits that put *H. floresiensis* high up the hominin tree, traits that link *H. floresiensis and H. erectus*—and there aren't many compelling ones. Given that *H. floresiensis* mostly exhibits primitive traits that put it close to the base of *Homo*, you'd be inclined to accept the evidence at face value and say 'That's where it belongs.'

> If you want to interpret those primitive traits as reversals and say *H. floresiensis* is descended from *H. erectus*, that's only possible if you find a whole suite of other characteristics that put it up there with *H. erectus*. You need overwhelming evidence that *H. floresiensis* is related to *H. erectus*, and I just don't see that.[43]

Mike here echoes the principle of parsimony, in which the most acceptable explanation for a phenomenon is the simplest. Scientists analyse and consider all possibilities and choose the solution that explains all the data, the solution that involves the fewest assumptions, and in our case, the solution that shows the fewest evolutionary changes. Cladistic analysis is based on this principle.

Another way to try to figure out the possible origins of *H. floresiensis* is to compare body and brain size differences among the likely ancestors. JAF Diniz-Filho (Universidade Federal de Goiás, Brazil) and colleagues assert that if *H. floresiensis* originated from an early *Homo*, such as *H. habilis* or *H. rudolfensis*, its small body and brain size does not need to be explained by the island rule and just reflects a deep ancestry.[44] That is, *H. floresiensis* could be explained by a very early, small-sized (and small-brained) species of *Homo* emanating from Africa and arriving on Flores with little evolutionary change.

What concerns Diniz-Filho and colleagues about this scenario is that it runs contrary to the Out of Africa 1 model. Because earlier forms of *Homo*, including *H. habilis* and *H. rudolfensis*, have never been found outside Africa, and they are at least 1.5 million years older than *H. floresiensis*, Diniz-Filho and colleagues opt for *H. erectus* as the ancestor of *H. floresiensis*, explaining that 'we feel it is not parsimonious to invoke an older and/or smaller-bodied African ancestor to explain *H. floresiensis* phenotype, especially if this implies revising the entire "Out of Africa I"'.[45]

But can we reject their first hypothesis, which proposes that a pre-*H. erectus* species diffused from Africa, simply because it does not agree with current dogma in human evolution—because it challenges the single Out of Africa 1 model of human evolution?

One of the roadblocks to the idea that the *H. floresiensis* lineage stems from a very early group of *Homo* is the lack of archaeological evidence for this outside Africa. The reasoning goes that, because there is no evidence for early hominins anywhere but Africa, then it is highly unlikely that any early, pre-*H. erectus* lineage would have left that continent. Interestingly, though, evidence has recently come to light that points to this scenario.

Assistant Professor Giancarlo Scardia (Universidade Estadual Paulista, Brazil) and colleagues discovered stone tools that date to between 2.5 and 2 million years ago in Jordan,[46] a period well before that espoused by the Out of Africa 1 model of human evolution. Two-million-year-old tools are known from much further east, too, in Pakistan and China.[47] These predate the recently discovered *H. erectus* in Africa by half a million years.[48] They predate *H. erectus* in Java by a million years or so, and they predate *H. georgicus* by 700 000 years. The only species known to have existed at the time of the tools discovered in Jordan are *P. boisei* and *H. habilis*, and possibly *A. africanus*. The Scardia and colleagues' finding might make it easier for critics to accept that something more primitive, more australopithecine-like, something like *H. floresiensis*, could have wandered out of Africa well before hominins arrived at Dmanisi.[49]

Still, we have no *skeletal* evidence for any early hominins anywhere between Africa and South-East Asia. And so, time and again, I hear this as an argument for rejecting *H. floresiensis* as an early hominin. It relies on the assumption that an absence of evidence can be a good reason for inferring that a species was never in a particular landscape. I find it perplexing that a case against an early hominin lineage spreading out of Africa is based on the lack of skeletal evidence anywhere between Africa and Indonesia, when we have no skeletal evidence for *H. erectus* between Africa and Indonesia either. Yet clearly *H. erectus* is on Java. Over thousands of years, *H. erectus* would have traversed the continents between Africa and South-East Asia. Logically, we'd expect the remains of at least some of these individuals to have been preserved as fossils. Yet no fossils of *H. erectus* have come to light across those regions. The absence of skeletal remains of either *H. erectus* or early hominins is more likely to be due to the fossils having not yet been discovered rather than anything else. It's a case of that old truism: absence of evidence is not evidence of absence.

I find it just as perplexing that the presence of *H. georgicus* from Georgia is invoked as an example of *H. erectus* moving out of Africa. If anything, this should alert us to the fact that small-brained species could move out of Africa: two of the *H. georgicus* brains are within the range of *H. habilis*,[50] and one is within the lower end of the range of *H. ergaster*.[51]

Now for some reflection. The implication of the work of my colleagues and I, and that of others, is that *H. floresiensis* is a remnant population of a lineage that emerged a long time ago in Africa. At some time, a group, or groups, of this lineage diffused from that continent to arrive at an unknown time on Flores. This dispersal represents a hitherto unknown movement of very early hominins out of Africa and challenges the Out of Africa 1 model of human evolution. Remember, too, that *H. floresiensis* survived on Flores until at least 53 000 years ago, more than a million years after its closest relative died out in Africa.

These are new concepts for us to grasp. They challenge us to rethink what we thought we knew about human evolution.

And there are other puzzles surrounding *H. floresiensis.* How did this species arrive on Flores, an island that emerged from the sea ten million years ago and that has never been joined to another landmass?[52] Fossil hominin remains of very small individuals have been discovered at the Mata Menge dig on Flores. What are they? Another diminutive species, *H. luzonensis*, was recently discovered in Callao Cave on the island of Luzon in the Philippines. What can this tell us about human evolution in South-East Asia? We will delve into these mysteries in the following chapters.

What's in a name?

There are strict rules for naming a new species of fauna or flora. This first came about in the eighteenth century when Swedish scientist Carl Linnaeus neatly streamlined the confusing botanical and zoological naming systems that had developed in an idiosyncratic way over many years. Under Linnaeus' system, a two-part name was applied to each species: the first part of the name is its genus and the second part is the species name. Hence, we are in the genus '*Homo*' and we belong to the species '*sapiens*'.

In assessing the characteristics of the newly discovered bones at Liang Bua cave, Peter Brown, Mike Morwood and colleagues found that the combination of features was unique, so much so that they proposed a new genus and species: *Sundanthropus floresianus*. However, before academic papers are published, experts in the same field of research extensively review them, and in this case the referees thought that the little Flores people, although unique in their combination of skeletal features, were not quite so different from everything else in the hominin family as to warrant a new genus.

The upshot of the peer review is that the new discovery is part of our genus *Homo,* but it is distinct from every other species in this genus. The authors proposed *Homo floresianus*. Luckily, one of the reviewers, who must have had a good knowledge of Latin, noted that '*floresianus*' means 'flowery anus'.[1] Imagine the fun generations of students would have had with that name! It was quickly amended to the more satisfactory *Homo floresiensis*, named for the island on which the bones were discovered.

4

Fossil bones of the So'a Basin, Flores

The ancestors of *Homo floresiensis*?

An absolute cacophony greeted Colin Groves, Dr Natasha Fijn (anthropologist and filmmaker, ANU), Professor Gregory Forth (anthropologist, University of Alberta) and me as we tumbled out of the Land Rover at the Mata Menge site in the So'a Basin. It was 2011 and we had been doing some archaeological fieldwork with our Indonesian colleagues, checking out possible sites for future excavation on Flores,[1] when Thomas Sutikna and colleagues invited us to visit the So'a excavations. Riveted by the noise, we turned as one towards its source, just as Gert van den Bergh and forensic anthropologist Kerrie Grant came towards us with welcoming smiles. Seeing the looks of amazement on our faces, Gert chuckled as he led us to the excavations. There, under an eerie blue light created by the sun shining through the tarpaulins above, more than a hundred trained local Ngada workers squatted side by side, pecking at or striking the solid rock with small chisels and hammers. We peered into the trenches and saw that massive bones were being exposed. Stegodont fossils! The sheer number of these astounded us. For Colin, it was an extra-special sight: stegodonts had always been of great academic interest to him. Quick as a flash, he'd donned his fossil-handling gloves, jumped into the trench and begun happily examining the bones as they were unearthed. I was content wielding a camera, not being au fait enough with stegodont bones to jump anywhere for them.

The So'a Basin on Flores has long been of interest to archaeologists and it continues to reveal intriguing finds—not just stegodont bones either, but stone tools as well. With the discovery of both at Mata Menge, archaeologists are now wondering just how early hominins arrived in the area and whether there might be a connection to *H. floresiensis*.

The 450-square-kilometre geological depression that is the So'a Basin lies 74 kilometres east of Liang Bua.[2] For much of its ancient life, the basin was a large lake or a series of lakes.[3] Today it is dry land with numerous small hillocks and occasional deep gorges, rich in plant life, covered in grass and

Gert van den Bergh, Kerrie Grant, Colin Groves (in white hat) and Ngada excavator (2011). Image provided by Debbie Argue.

dense scrub, and flanked by volcanoes. It is named after the small town of So'a on the western fringe of the basin.[4] The area became a focus of research in the 1950s when the discovery of stegodont bones allegedly associated with stone tools stretched credulity. Stegodonts emerged during the Late Miocene and lived until the Late Pleistocene across Asia, including Indonesia and the Philippines, and in East and Central Africa. In Java they are known from at least 1.2 million years ago. Could hominins really have been on Flores in the era of the stegodonts?

The first documented discovery in the So'a area was back in 1956. During a hunting trip in December that year, the Raja of Nage-Keo at the district capital Bo'a Wae, Yoseph Juwa Dobe Ngole, discovered large bones exposed by erosion near the abandoned village of Ola Bula.[5] By chance, a Catholic priest, Father Theo Verhoeven, lived nearby. He had a master's degree in classical language and literature that included some archaeology, and he retained an abiding passion for the latter subject. A friend of the Raja, Castillio, showed Verhoeven the fossils found by him. In February 1957 Verhoeven visited Ola Bula, initially alone because Castillio was ill, but later accompanied by Castillio, and they collected more fossils.[6] The bones certainly looked interesting and, in April 1957, together with two members of the museum at Bogor, Java, Verhoeven excavated the site and collected bones from the surface. The Raja agreed to send the collection to a vertebrate fossil expert at the Natural History Museum in Leiden, Dr Dirk Hooijer, who was *the* expert in Indonesian fossils and those from South-East Asia (he would be the curator at the museum from 1946 until 1979).[7] Dr Hooijer identified the bones as stegodont and named them *Stegodon trigonocephalus florensis*, a subspecies of a stegodont species known from the fossil record of Java.[8]

Verhoeven also collected stones from the site that showed signs of human manufacture. The geologist HMS Hartono found similar stones during the Geological Survey of Indonesia's mapping project in the region in 1960. Hartono remarks in his typewritten report of the geological investigations that the 'surface scatter bone fragments are usually found together with "artifacts" and boulders varying in size from 3 cm to 5 cm in diameter'.[9]

The stegodont situation became more complex when Hartono found two milk, or infant, molars belonging to a stegodont, but which were half the size of the Java stegodonts (*S. trigonocephalus*). Hooijer studied the molars and attributed them to a pygmy stegodont.[10] In later years, when much more material became available, the pygmy stegodont bones were thoroughly studied by Gert van den Bergh and attributed to a new species, *Stegodon sondaari*,[11] named after Dr Paul Sondaar in recognition of his achievements as a scientist and his contribution to student learning—Dr Sondaar was a vertebrate fossil expert with a longstanding interest in the evolution of animals on islands. However, just how old the So'a Basin infant stegodont teeth were could not be established because they had been found on the surface of the ground and it was not possible to know from where they had eroded out.

By the 1960s, then, we had a large stegodont, a pygmy stegodont and stone tools on Flores, but no idea as to if, or how, they might be linked.

Verhoeven returned to the So'a Basin in 1963, where he did some small excavations at a new locality, Bo'a Leza, some 4 kilometres from where the first stegodont bones had been found. The workmen quickly came upon a disconnected skeleton of a stegodont. Its best-preserved tusk was 2.8 metres long—not a pygmy skeleton then. The teeth of crocodiles and rodents were also found. Of particular importance, stone tools were found within the stegodont fossil beds.[12] Two years later, Verhoeven made a small excavation at another locality in the So'a Basin: Mata Menge. Once again, he found stone artefacts with fossil stegodont bones.

By 1965, then, based on the evidence of the stone tools, it was looking very much like early hominins had been on Flores at the same time as stegodonts. It seemed astonishing that hominins could have been around at this time on Flores, an island that had never been attached to a mainland. The general view back then was that only our species, *H. sapiens*, had the intellectual and technological capacity to make sea crossings, not early hominins.

Verhoeven invited Father Johannes Maringer, who had taught archaeology at the Catholic University in Japan,[13] to join him on an archaeological expedition on Flores. The pair headed to the So'a Basin and confirmed that the artefacts and bones there occurred in the same layers, not only at Mata Menge but at a number of other localities as well. They concluded:

> The presence of Pleistocene man in Flores is proved by the association of stone artefacts together with the bones of *Stegodon trigonocephalus floresensis* ... The known hominids from Java range from the Middle Pleistocene *Pithecanthropus erectus* [now *H. erectus*] and the ... Upper Pleistocene *H. soloensis* ... It is to be hoped that future excavations in the Mengeruda and Old Bula areas will ... above all, discover human skeletal remains, which would enable us to determine the kind of hominids which first entered Flores.[14]

As is often the way in the world of archaeology, Maringer and Verhoeven's claims tended to be dismissed by archaeologists. There were questions over whether the stones described as artefacts were indeed artefacts, and whether they and the stegodont bones really did occur together.[15]

A couple of years later, after Verhoeven had retired to Belgium, Dr Paul Sondaar fell into conversation with him.[16] Sondaar was intrigued by what Verhoeven had said about the stegodont bones, especially the pygmy ones, and the stone tools on Flores. In fact, he was so intrigued that he headed off to Flores in 1980 in the company of the late Professor Sartono (geologist and palaeoanthropologist, Institute of Technology Bandung), Professor Soejono, Rokus Awe Due and Swiss palaeontologist Burkart Engesser. They visited a new locality, Tangi Talo, 4.5 kilometres east of Mata Menge. Here, Sondaar's team discovered the remains of pygmy stegodont bones and teeth, and giant tortoise fragments,[17] but not stone tools.

The team also took the opportunity to explore the So'a Basin, looking for other sites, and it became clear that this area had huge archaeological potential, worthy of a much larger research project. Sondaar's PhD student Gert van den Bergh was invited to join Sondaar and de Vos in that undertaking, but it would be ten years before funding became available for further research.

When the team finally headed back to Flores, they had an ambitious agenda: check out how the sites all related to each other in chronology; check the claims of Verhoeven concerning the association of stone tools and stegodont fossils at Mata Menge and other locations; collect more fossil material, particularly of the pygmy stegodont; and sample the various sections at the archaeological sites to date them.[18] Gert recalls: 'I went to the So'a Basin in April 1991. That was with pak Yep [geologist], with the white hat in the photo, [opposite, bottom] pak Suyatman [geologist] and pak Mukid [technician; all three from the then Geological Research and Development Centre, Bandung] with me. Aziz could not join the first time.'[19] No-one in the team had been to the Mata Menge site, so relocating it in the vast So'a Basin would, they thought, be a challenge. But, says Gert, it was 'a piece of cake. The local villagers knew exactly where it was!'[20]

Later, Paul Sondaar's team, including Fachroel Aziz, Iwan Kurniawan,[21] John de Vos and Gert van den Bergh, dated samples from a section of Verhoeven's 1970s Mata Menge excavation and the Tangi Talo site.[22] To do this, they used a technique called palaeomagnetic dating.

Strange as it might seem, the Earth's magnetic field has swung from north to south and south to north many times over the eons. When rocks form, the magnetic minerals orient themselves to the magnetic field in existence at that time and get locked into that position. Archaeologists can extract cores of rock from the strata in which they are interested and send them to a laboratory where special instruments are used to measure the orientation of the magnetic particles in the rock.[23] Scientists have worked out when each 'swing' took place. The latest transition of the magnetic poles from south to north, or 'normal' polarity, was around 780 000 years ago; this event is called the Brunhes-Matuyama boundary and marks the beginning of the Middle Pleistocene.[24] The other thing to note is that palaeomagnetic dating is a 'before' or 'after' technique, as opposed to obtaining an absolute date for tools, bones or other items of interest in your excavation.

Dr Bondan Mubroto performed the palaeomagnetic dating of the samples from Tangi Talo and Mata Menge. The Tangi Talo fossil layer had 'reverse' polarity. The Earth's magnetism was last oriented to the south 780 000 years ago, unlike today when it is oriented to the north. The Tangi Talo fossil-bearing layers were therefore older than 780 000 years, and they were also older than the Mata Menge tool- and fossil-bearing strata, which had normal polarity. Tangi Talo was more recently dated to 1.4 million years ago.[25]

Top: The Indonesian–Dutch team in 1992. On the far left is the late Bondan Mubroto, who did the palaeomagnetic analysis. Third from the left in the back row is Iwan Kurniawan; fifth from the left (standing) is the late Musa Bali, who organised the local workers. The late Paul Sondaar is third from the right (in the white shirt). To his right is pak To, the owner and driver of the vehicle. Image provided by Dr John de Vos. Identities of individuals supplied by Gert van den Bergh.

Bottom: The team in 1994: Musa Bali (left); pak To (second from left); Slamat Sudjarwadi (technician, Museum Geologi, third from left); John de Vos (fourth from left); Gert van den Bergh (fifth from right); pak Minggus (traditional elder and top fossil hunter, fourth from right); and Fachroel Aziz (second from right). Image provided by Gert van den Bergh.

Because the Mata Menge site had normal polarity,[26] the team inferred the site to be less than 780 000 years old, just as Maringer and Verhoeven had predicted. The artefact-bearing layer at Mata Menge was 3 metres above the Brunhes-Matuyama boundary, so Sondaar and colleagues surmised that this site was a little younger than the reversal time.

Paul Sondaar and others speculated that the major differences between the Tangi Talo and Mata Menge sites resulted from the arrival of *H. erectus*, who subsequently hunted the pygmy stegodont and giant tortoises to extinction.[27] It is, after all, what we modern humans seem to do time and time again: arrive in a new locale and wreak havoc on the landscape, including destroying species.

Based on the discoveries of stone tools and animal bones, and the dating process, let's explore the idea of *H. erectus* as the toolmakers.

Although fossils of *H. erectus* have never been found on Flores, it is almost a given that *H. erectus* was the species on the island that made the tools, because that was the only early hominin species known in the region. It lived not far away, on Java, from 1.5 million or more years ago (see Appendix A). There is just one more thing to add here: the astonishing implication is that *H. erectus* must have been able to cross water barriers.

We had never considered this as something hominins could do. We had thought that only we modern humans had the mental capacity and drive to risk going over water to get somewhere.[28] So the concept that hominins could make water crossings was a remarkable one in human evolutionary terms. However, two things hampered the acceptance of the team's pronouncements. Not only were they made in somewhat obscure publications with a limited audience,[29] but, says Gert, 'because we are mostly geologists and there were no stone tool experts involved, not much notice was taken.' Those who did take notice questioned the relatively old age of the artefact-bearing layers and whether it was *H. erectus* who'd made the tools, or they stressed that it was not known when the stegodonts became extinct.

The resolution of the issue of the artefacts as tools and *H. erectus* as the maker was critical in assessing if water crossings could have been made. Fachroel Aziz thought it would be useful to ask a recognised stone tool expert to check over the Mata Menge stones to confirm whether or not they were artefacts. The stones were being kept at the Geological Survey Institute in Bandung, and by chance, Mike Morwood happened to be in Jakarta at the time, only a couple of hours away. Acting on Aziz's request, Morwood hastily made for Bandung and was able to verify that most of the stones were just as had been recognised: stone artefacts.

This spurred Mike and his team into action. Because the palaeomagnetic dating of the Mata Menge and Tangi Talo sites did not date the fossil deposits themselves, the team wanted to find out the age of the layers in which the archaeological finds had been excavated. So they sent mineral samples from the dig sediment to a dating lab. The resulting analysis showed that the Tangi Talo site was 900 000 years old and the Mata Menge site was 800 000–880 000 years old. Publishing these results in the prestigious scientific journal *Nature*, Morwood and colleagues argued that *H. erectus* had reached Flores by 880 000 years ago.[30]

I remember being astounded by the idea that *H. erectus* could make water crossings, and I asked Colin Groves if that really was the implication of Morwood and colleagues' discovery. He replied in the affirmative.

The *Nature* article served as the basis for a successful grant application to the ARC and ongoing support from the Geological Survey Institute. This heralded a new era of research. It seemed that the huge archaeological potential identified by Paul Sondaar all those years ago was finally being realised.

Large-scale excavations have now been initiated at four sites. Sixteen stegodont fossil sites have been located and the geology of the entire So'a Basin has been mapped.[31] At Mata Menge alone, a dozen archaeologists, geologists and palaeontologists, predominantly from Indonesia and Australia, are involved. Over a hundred local people have been trained in excavation procedures and are employed on-site and in sorting bones and stone tools. Visiting specialist collaborators from New Zealand, the Netherlands, and Denmark contribute to the research program. In 2013, Gert van den Bergh and his team commenced a new excavation about 13 metres higher up in the Mata Menge sequence. It is perhaps as much as 80 000 years younger than the lower level. By 2019, they had excavated 956 square metres at Mata Menge alone (718 square metres in the lower level and 238 square metres

Top: Mata Menge towards the end of the excavation in July 2012. Image taken by Kerrie Grant and supplied by Gert van den Bergh.

Left: Excavations at Mata Menge in October 2013. Image provided by Gert van den Bergh.

in the upper level), while also having excavated other sites in the So'a Basin, such as Tangi Talo, Wolo Sege, Bo'a Leza and Malahuma.

The archaeological wonderland doesn't end there. About a 500-metre walk from Mata Menge is a site called Wolo Sege, where archaeologist Adam Brumm (now a Professor at Griffith University) and his team made an interesting discovery. It has become a deep excavation, but when I visited it in 2011, I wondered why Adam had chosen this place to dig when the entire landscape seemed unremarkable, archaeologically speaking. Here's Adam talking about the chance experience that led to the establishment of the site:

> I was part way through my PhD in studying stone tool technology. We were excavating at Mata Menge in 2005. During lunch breaks I liked to wander around. It was blindingly hot and on one occasion I became lost, disorientated. Stumbling down little valleys I was trying to find out where I was. I came down into this little buffalo yard. People created these *kendang kerbau* usually at low points in a streambed, where a natural hollow had formed. In this case the hollow had a steep-sided embankment—a natural enclosure for buffaloes, which people closed off at one end with a fence. I could see that the embankment had lots of nicely stratified layers exposed so I walked into it and found these huge stone tools lying around on the surface of the stock yard: huge chopping tools and flakes. And looking at the wall of the cattle yard I saw layers where these might have been eroding from.[32]

Adam put two and two together. He knew that the strata in the So'a Basin were clearly layered. So the lower you were in the strata, the older would be the artefacts.

> The fluvial strata from which I guessed the artefacts had eroded ... were really low in the stratigraphic sequence exposed in the rock wall of the *kendang kerbau*. I managed to find my way back to Mata Menge and brought Gert [van den Bergh], whose knowledge of the basin geology was much more developed than mine, back to the site. The real significance of Mata Menge at the time was that it was thought to be the oldest site on Flores with stone tools—the oldest site where hominins had been. But it looked like we could have much older stone tools at this new site Wolo Sege. Could it be that hominins got to Flores earlier than we'd thought? This was the idea running through our minds as we assessed the potential of the site.[33]

Bringing over several members of the Mata Menge fieldwork team, Adam set to work. At the base of the excavations they found stone tools, and just above the tools was a layer of volcanic material from an eruption. When they dated samples from this volcanic layer, it turned out to be just over a million years old.[34]

A couple of questions arose from Adam's work. One concerned the culpability of the hominins in the demise of the pygmy stegodonts. The other was: who were the hominins?

Regarding the stegodonts, well, the hominins are off the hook. Back at the Geological Museum lab, Gert van den Bergh carefully analysed the bones from Mata Menge. His analysis established that the stegodont bones represented individuals of all ages: adults, young adults and juveniles. The frequency of each age category was in the same proportion as that of a living herd: 30 per cent juveniles, a few teenagers and so forth. Had the stegodont bones in the Mata Menge excavations been primarily of one age group, we could have inferred hunting by hominins as the cause of death.

Adam Brumm showing Colin Groves the one-million-year-old layer in the Wolo Sege excavation (2011). Image provided by Debbie Argue.

We saw this in the Liang Bua excavations, where over 90 per cent of the stegodont bones were from juveniles,[35] but this was not the case for Mata Menge. Van den Bergh and colleagues reasoned from the age range represented by the bones, and details of the sediment at the site, that a sudden and catastrophic event had killed an entire local population of stegodonts. The bones were exposed to the elements and transported by streams before they were buried by a series of mudflows.[36] So hominins had not caused the extinctions. It was largely coincidental that stone tools occurred along with the fossil bones.[37]

As to the tantalising question of who the hominins were, it had never really been questioned that it was *H. erectus* who'd made it to Flores a million or so years ago. It made sense. As mentioned above, the species had lived not too far away, in Java, and until 2004, when *H. floresiensis* was announced, it was the only hominin species known in the South-East Asian region. But something new was about to bounce onto the scene.

In January 2015, Gert van den Bergh and his partner Dr Susan Hayes (University of Wollongong), in the midst of a road trip, stayed with me and my husband in Canberra. Colin Groves and his wife Phyll joined us for dinner. Strolling into the kitchen, Gert casually mentioned that there was something he wanted to show Colin and me. I thought Gert was going to present the latest photos from his excavations in the So'a Basin, but instead he brought out a little box, and inside that was tissue paper, and inside the tissue paper was ... a cast of a partial hominin jaw. The jaw was tiny, much smaller than the *H. floresiensis* jaws; it nestled easily in the palm of my hand. Also in the box were some small adult teeth and the tooth of a child. I think I said something like, 'What's this!?', when I could, of course, see what everything was. But what did it all mean?

It meant that Gert's team had discovered the long-wished-for fossil hominin remains at Mata Menge. This was the prize we'd coveted ever since Father Verhoeven had discovered the stone tools back in the 1960s.

Gert recently described to me how the fossils were found:

> It was towards the end of the final year of our research project, in October 2014, so Iwan [Kurniawan] and I were trying to find new sites to give us

> a better chance for more funding. With speleologists from Yogyakarta we were surveying caves to the north and west of the So'a Basin. We were in an isolated place and there was no phone signal. In the meantime the Mata Menge excavations continued. It was only when we got back that we heard the news. Andreas [Boko Rema], one of the local village assistants, had excavated a small tooth. He came to Mika Puspaningrum, my PhD student who was trench manager (a manager is assigned to each of the five trenches) to check it. Mika recognized it as hominin. We were very excited when we saw it. The next day we excavated an incisor, a premolar, a piece of mandible, and two days later, a milk canine. We had this enormous party with the local villagers, music and there was dancing. It was marvellous.[38]

It was also unexpected, as Gert explained to Ewen Callaway, a senior reporter for *Nature*:

> ... because the layers where we were digging are about 600 000 years older than *H. floresiensis*. We thought that the hypothesis that *H. floresiensis* was a dwarfed form of *H. erectus* is the most likely one, so we expected to find in those early layers big-bodied *H. erectus*-like species but instead what we've found are fossils very similar to *H. floresiensis*.[39]

Gert was referring here to one of the two proposed explanations for *H. floresiensis*: in this case, that the species dwarfed from an unknown population of *H. erectus* that arrived on Flores, as an evolutionary response to the particular conditions encountered on the island. The early dates for Mata Menge led to the team's expectation that *H. erectus* might be discovered here.

In writing about the tiny bones in *Nature*, the team's central idea was that *H. erectus* arrived on Flores at an unknown time and dwarfed into the Mata Menge hominins, which were ancestral to *H. floresiensis*. The research team was cautious, however, about formally attributing the bones to *H. floresiensis*, calling the fossils '*floresiensis*-like'. In the team's words: 'We conclude that the most reasonable taxonomic assignment for the Mata Menge fossils is to

H. floresiensis, although this remains a provisional interpretation until new skeletal materials are found.'[40]

The layer from which the remains were excavated was dated to between 650 000 and 800 000 years ago.[41] As Gert says, that makes them at least 660 000 years older than *H. floresiensis*. Stone tools and fossils of pygmy stegodonts, rats and komodo dragons were found in the same level as the fossil hominin bones.

But back to the bones themselves. In total, a partial jaw, six teeth, and a piece of bone that could be part of a hominin skull, were found within 15 metres of each other in the excavation, representing at least three individuals: one adult and two infants. These fossils document the existence of a population of very small hominins living on Flores by at least 800 000 years ago.[42]

To explore whether there is any possible link with *H. floresiensis*, let's take a look at the Mata Menge bones. The first thing to note is the size. This little adult jaw is 21–28 per cent lower and narrower than the two *H. floresiensis* jaws from Liang Bua cave.[43] The team explained the increase in the size of the Mata Menge hominins to the size of *H. floresiensis* as reflecting either a literal increase in size in individuals over time, or variation in size among the Mata Menge and Liang Bua hominin groups. There was certainly variation among the individuals of *H. floresiensis*. In chapter 1, we saw that Bill Jungers and colleagues noted that the partial skeleton LB1 appears to be the largest individual recovered for *H. floresiensis*.

When the Mata Menge fossils were announced, some of my colleagues wondered if it was the jaw of a child. It would appear not: the team's CT scans showed the root cavity where there was once a third molar, which is an indicator of adulthood. The third molar had disappeared from the jaw some time after the individual died, leaving just the hole in which it had been embedded.[44]

The Mata Menge jaw is fragmentary. It comprises only part of one side, the right. Missing is the left side: the front part (chin area), so we cannot see if the jaw had those internal buttressing structures that *H. floresiensis* jaws had, or if it was internally more like *H. erectus*' jaws (or those of any other hominin species). Also missing is the ramus or bony part that articulates with the skull. There are, then, only a limited number of observations and comparisons that can be made to help clarify its origins and to assess if it is the ancestor of *H. floresiensis*.

Van den Bergh and colleagues found that the Mata Menge jaw had features that became apparent in post-1.7-million-year-old hominins, including *H. erectus* and *H. floresiensis*. This suggested that early hominins such as *H. habilis* were not involved in the ancestry of the Mata Menge hominins. They also described specific similarities between the Mata Menge jaw and *H. floresiensis*, and noted that the former lacked the characteristics of *Australopithecus* and *H. habilis* jaws. This supported the idea that Mata Menge was not too closely related to *H. habilis* or the australopithecines.

The breadth and height of the Mata Menge jaw could be measured in a couple of places, which provided a cross-sectional snapshot of the jaw shape. The researchers were able to compare the height/breadth of the Mata Menge individual with that of the jaws of *A. afarensis*, *H. habilis*, *H. erectus*, *H. floresiensis* and the Ledi-Geraru jaw. Their results showed that the Mata Menge jaw, in its height-to-breadth ratio, was virtually indistinguishable from the 2.8–2.75-million-year-old Ledi-Geraru jaw (see Appendix A) and one of the three, chronologically younger, *H. erectus*/*Homo pekinensis* jaws.[45]

As well, there were two jaws of *H. erectus* and two of *H. habilis* in the analysis, and the Mata Menge jaw differed from each of these. Notably, in the aspect of jaw shape, Mata Menge (and Ledi-Geraru and *H. erectus*/*H. pekinensis*) differed from the two *H. floresiensis* individuals. In fact, the *H. floresiensis* LB6 jaw was very similar to an *A. afarensis* jaw in its height/width proportions. Yet, as we saw above, *H. floresiensis* and Mata Menge have several characteristics in common that are not shared with the australopithecines.

The team viewed the jaw analysis differently: they saw Mata Menge as similar to *H. floresiensis* in shape because the Mata Menge jaw 'occupies the space in between the two Liang Bua *H. floresiensis* mandibles, suggesting their shared lateral corporal shape'.[46] That is, they saw the Mata Menge and *H. floresiensis* jaws as similar in their height-to-width proportions.

Six hominin teeth have been recovered from the Mata Menge excavations. The bevelled wear pattern on the incisor and premolar suggests that the Mata Menge hominins had substantial prognathism, a primitive condition found on the australopithecines, *H. habilis*, *H. georgicus*, *H. ergaster*, *H. floresiensis* and *H. erectus*. The form of the root on the Mata Menge premolar and the types of crests it has also exemplify primitiveness, as do the forms of the crests on the molar.[47]

Unfortunately, as the researchers noted, most of the Mata Menge teeth are not informative when it comes to working out to what species this population might have been related. They are either too fragmentary or worn (the lower incisors), or we don't have the equivalent teeth for *H. floresiensis* (deciduous teeth), or the tooth in question does not vary among *H. habilis* and *H. erectus* (upper incisor), or the features on the teeth are not useful in resolving the question of Mata Menge relationships (upper incisor and upper premolar).[48] That leaves just a lower first molar.

Van den Bergh and colleagues applied a number of analyses to see how the Mata Menge molar compared to the key species of *H. floresiensis*, *H. erectus* and *H. habilis*. The first thing we see is that the Mata Menge molar is moderately long while the *H. floresiensis* molars are wider than they are long,[49] as noted by the researchers.[50] Although long, the Mata Menge molar does not have the more extreme length that is found in some *H. habilis* molars, suggesting to the researchers that there was a significant difference between these two groups of hominins.[51]

The other aspect of the molar that can be used to help clarify possible relationships with other species is the shape of its crown contour. In this, the Mata Menge molar is similar to Java *H. erectus* (and ours).[52] Interestingly, here we can see something unexpected about *H. floresiensis*: its molar contours are distinctly different from anything in the analysis. They are unlike Mata Menge, *H. habilis*, *H. erectus* or modern humans.[53]

In sum, the Mata Menge remains are not giving us a clear-cut signal as to which species they are most closely related to, or how they might relate to *H. floresiensis* or *H. erectus*.

The problem that faces us here is an inescapable one: the fossil material from Mata Menge is fragmentary and there is not much of it, not yet—it's not enough on which to base hypotheses. We really need more of this jaw. In fact, as the team has pointed out, we need more bones altogether. This is especially so because the characteristics telling us that *H. floresiensis* derives from a primitive ancestor, come from the skull, shoulder, wrists, arms, legs and feet, and the rest of the jaw. It will not be until more fossils from Mata Menge are discovered that we can assess where they fit in the human evolutionary story.

So where does that leave us? At this point I believe that the evidence for any ancestor–descendant relationship between Mata Menge and *H. floresiensis* seems ambiguous. Also ambiguous is the evidence for Mata Menge dwarfing from *H. erectus*. While I feel a certain sense of discomfort that I'm not quite on the same page as my friends Gert and team, I know that science progresses by questioning hypotheses or ideas, so dissension is not such a bad thing. All things change—hypotheses or ideas may later be corroborated, or they might require revisions or refinement, or simply be superseded when new evidence emerges. Until then, we need to keep an open mind about this exciting, enigmatic population of tiny hominins.

And who knows? It could even be that we have a new little species here.

5

Float, walk or swim?

How did *Homo floresiensis* get to Flores?

'The Thorpe factor!' said my friend Julie as I described *H. floresiensis*' long feet. 'Maybe it swam to Flores?'

Julie was of course referring to the famously flipper-sized feet of Australian Olympic gold medal swimmer Ian Thorpe. At size 17, Thorpe's huge and super-flexible feet (apparently he could touch his shin with his toes) along with his six-beat kick gave him a huge boost over his rivals in the pool in the early 2000s. But what really got my attention was that Julie, with this comment she made in early 2016, had inadvertently hit upon a question that had long puzzled me: could hominins somehow have made sea crossings? I wanted to explore this idea in more depth, drawing on various archaeological discoveries in Indonesia over the years, as well as natural occurrences like tsunamis, cyclones, other weather patterns and ocean currents.

Archaeologists and human evolution researchers generally take it for granted that, of the hominins, only modern humans had the brainpower and technological ability to make sea crossings. The earliest evidence for this accomplishment was the arrival of modern humans in Australia between 40 000 and 60 000 years ago. In my experience, to human evolutionists, the idea that early hominins possessed the ability to cross water, through the use of seagoing craft or, especially, a natural talent for swimming, is like a red rag to a bull. Yet archaeological work on Flores contradicts this strongly held view. As Gert van den Bergh says: 'The unexpected conclusion looms up: hominins were able to make sea crossings. If this conclusion will stand critical evaluation ... it will alter the image of our direct ancestors.'[1]

When *H. floresiensis* burst onto the scene in 2004, researchers started asking just how this hominin could have made it to Flores. Mike Morwood and colleagues floated two ideas in the 2009 edition of the *Journal of Human Evolution* that was dedicated to studies on *H. floresiensis*. One of their ideas was that *H. floresiensis* could have arrived from Sumbawa, which lies to the west of Flores. The other was that Sulawesi, a large island to the north of Flores, could have been the place of origin. Morwood and colleagues favoured the Sulawesi proposal because of the Makassar Strait Throughflow,[2] a strong current that flows south from the North Pacific Ocean through the Makassar Strait between Borneo and Sulawesi, and into the Indian Ocean—with some branching along the way. The researchers argued that, as a result of an extremely rare event such as a tsunami, a small colonising group could have accidentally arrived on Flores on this current while clinging to a natural raft of vegetation or an uprooted tree that had been washed out to sea.[3]

Robin Dennell and colleagues also proposed that *H. floresiensis* or its predecessor was likely to have arrived on Flores by floating on natural rafts of vegetation dislodged during cyclones or tsunamis.[4] They, like Morwood and colleagues, envisaged two possible scenarios. One was that hominins had been swept off western Sulawesi following a tsunami and got caught up in the throughflow, eventually arriving on Flores.[5] Their alternative suggestion was that *H. floresiensis* dispersed via Java, Bali, Lombok and Sumbawa.

The idea of hominins being accidentally transported to Flores via a powerful current continued to gain momentum, and by the time I became interested in exploring the idea, it had become the general consensus. But in February 2016, when I explained to oceanographer Dr Stewart Fallon on the ANU campus that I was keen to find out how long it might take for hominins to float down the Makassar Strait Throughflow to Flores, he looked somewhat shocked. I was momentarily perplexed: had I simply asked too much of this busy academic? Then Stewart's face relaxed into an amused smile and he said, 'The Indonesian throughflow flows 200 metres under water.' He brought me out of my stunned silence by saying that it was the surface currents I'd be interested in. 'The surface currents are driven by the monsoon winds. Look ...' he said, uploading maps to his computer screen. I could see that during one of the two monsoon seasons, the currents would flow south, and during the other monsoon season they would flow north. I left Stewart's office clutching a list of articles to track down.

My reading confirmed that while the throughflow does indeed enter the Makassar Strait as a surface current, it then plunges to become a strong, jet-like current at a depth of between 70 and 240 metres, eventually splitting when it reaches the Java Sea.[6] One branch heads into the Flores Sea. As it passes further south through the Lombok Strait there is a surface component to the current, but the main core of the flow is subsurface. The other exit passages are in the Ombai Strait and Timor Passage,[7] where the water moves independently of surface flows.[8] The Indonesian throughflow current simply couldn't have taken hominins anywhere from Sulawesi.

Could surface currents driven by monsoon winds have propelled hominin tsunami victims from Sulawesi to Flores? More research revealed that, from January to March, winds push currents in the Makassar Strait southwards into the Java Sea, while from July to September, the currents flow eastwards from the Banda Sea, northwards into the Makassar Strait and westwards into the Java Sea.[9] It is only the southward-flowing currents that could potentially have dispersed hominins from Sulawesi in the direction of Flores.

The question now was whether monsoonal currents flowed in these same directions a million years ago, when early hominins were getting to Flores. At first this seems like an impossible question to answer: no modern humans were around at that time to bear witness to what was happening. There are records, though, which have resulted from the scientific examination of microscopic plant pollen remains and other soil components from archaeological excavations. From these, it has been possible to figure out ancient climate conditions. For the Sangiran region of Java, for example, where the 1–1.5-million-year-old *H. erectus* fossils were discovered, Brasseur and colleagues found that the climate back then was just as it is today: mainly influenced by South-East Asian monsoon cycles.[10] Studies in China by T Liu and Z Ding found the same thing, except that their efforts revealed that consistent monsoon weather patterns go back at least 2.5 million years, although they varied in intensity from time to time.[11]

We can assume, then, that monsoons, their attendant winds, and the currents those winds generated, were occurring when the ancestors of *H. floresiensis* reached Flores. So let's look more closely at how floating to Flores could have panned out.

As mentioned above, Morwood and Dennell and their respective colleagues suggested that hominins could have floated to Flores by clinging to a raft comprising vegetation that was dislodged by a tsunami-like event. Dean Falk suggested an alternative scenario: perhaps, based on a foot morphology that indicates partial arboreal habits, *H. floresiensis* slept on tree platforms like orangutans. If so, they may have nested in groups each evening. Imagine them gathering in trees on the outside curve of a Sulawesi river swollen in the wet season. Big flows might have eroded the banks, with occasional large chunks falling into the water. If such a chunk had multiple trees with nesting individuals, this could have been the large, tangled natural flotsam on which *H. floresiensis* was swept away.[12]

The three primary elements in this scenario are a tsunami (or storm) event, the formation of a natural raft of vegetation, and surface flows. We also need to consider the probability of early hominins surviving the voyage to land on Flores and going on to establish a viable population.

Tsunamis are caused by volcanic activity that affects the ocean floor and generates a powerful surge of water. As a tsunami approaches land, it appears as a violent onrushing tide. The water movement can be very complex, with successive waves becoming larger and larger and often colliding with each other. These waves are faster, higher and contain much more energy than cyclonic storm surges. They may inundate hundreds or even thousands of metres of inland terrain, and the onslaught of successive waves can last for many hours. The toll in injury and loss of life, both onshore and as people are washed out to sea, can be catastrophic.

We have witnessed this level of destruction in recent times. On Boxing Day 2004, a 30-metre mountain of water hit Aceh, a densely populated coastal area of Sumatra characterised by many thousands of settled communities of fisherfolk and farmers. The tsunami killed 100 000 men, women and children. Buildings folded like collapsing stacks of cards, while trees and cars were swept up in the oil-black rapids—very few of those people caught up in the deluge survived.[13]

In 2006, another tsunami hit a 300-kilometre coastal stretch of southern Java, with most of the damage centred on the town of Pangandaran. According to witness accounts, the sea level fell before a 3-metre wave struck the shoreline; the next wave was 5 metres high. The water surged 500 metres inland.[14] Some 525 people died and 273 were unaccounted for; more than 50 000 people were evacuated.

The 2018 Sunda Strait tsunami claimed the lives of at least 426 people and injured over 14 000 others. The Indonesian Navy stated that dozens of bodies were recovered from the sea.[15]

Hamzah and colleagues compiled data from several sources and found that 105 tsunamis occurred in the Indonesia region from 1600 to 1999.[16] Nine of these originated in the Makassar Strait, sixteen occurred off Sumatra, and nine were in the eastern Sunda Arc, an area that includes Java, Bali, Lombok and Sumbawa. There is no apparent pattern of tsunami occurrence. For example, the earliest tsunami listed in the researchers' compilation for the eastern Sunda Arc, a region we consider below, occurred in the decade 1810–19. Another tsunami occurred in this region in the 1850s, followed by three in the next decade. Then there was a forty-year gap until the next one (in the period 1910–19); fifty years later, in the 1990s, came the last tsunami of the twentieth century. We know from this that there can be long intervals between tsunamis, but they are just as likely to occur in quick succession. Of course, we cannot know the frequency of tsunamis before written records or oral histories were kept—we have no indication of the frequency of tsunamis that took place a million years ago.

The next important element of our floating-to-Flores scenario is the mode of transportation. Natural mats of vegetation comprise either small sections of land that have detached from the coast or a riverbank, or tangles of plants, trees and roots that have been dislodged and entered the water.[17] Perhaps surprisingly, little is known about the number or frequency of such dispersals of animals through time—we don't even know much about modern-era events, with the available evidence to a large extent being circumstantial.[18] What we do know is that, typically, it is insects and reptiles that are caught in this way. Theil and colleagues noted that episodic events occur at intervals of many years, often decades or even centuries apart.[19] The few accounts of animals found on rafts mainly involve single individuals, although Censky and colleagues described the arrival of at least fifteen iguanas on the eastern beaches of Anguilla in the Caribbean a month after two successive category 4 hurricanes moved through the region—they arrived on a mat of logs and uprooted trees 9 metres long.[20]

It would seem like a relatively straightforward notion that surface currents in the Makassar Strait might have dispersed hominins to Flores, but it is not that simple. The currents are complex systems that include eddies, which are circular or elliptical ocean currents that form at the southern entrance of the Makassar Strait, and which can vary from tens of kilometres to hundreds of kilometres wide. Although we do not know if eddies occurred in the strait a million years ago, I think it is worth keeping this possibility in mind.

Nuzula and colleagues recorded twenty-nine eddy events in the Makassar Ṣtrait between 2008 and 2012; these were between 124 and 255 kilometres wide. Eddies may swirl adjacent to coastal areas or further out to sea at any time of year, with their frequency varying from weekly to monthly. For example, in January 2009, a 132-kilometre-wide eddy developed adjacent to the south-western shoreline of Sulawesi. It circulated in a clockwise direction and possibly struck the coast.[21] The following month, a 147-kilometre-wide anticlockwise eddy formed that abutted the western coast of Sulawesi.[22] The horror scenario for anyone swept off Sulawesi by a tsunami is to be trapped in an eddy, where there are no means of escape.

Stories of the occasional human survivor of an Indonesian tsunami are sometimes used to illustrate, or back up, the possibility that island colonisation can happen as a result of tidal wave action.[23] One of these concerns a pregnant woman who was rescued from a floating sago tree five days after the 2004 Aceh tsunami.[24] The woman's survival may sound like compelling evidence for possible island colonisation, but what was not reported in the associated research was that she could not swim and had been thrashing about, trying to keep her head clear of the water, when she chanced upon a tree trunk. She survived by eating the fruit and bark of the sago palm. The fact that she twice slipped from the tree but managed to hold on, and saw sharks around her in the water, draws further attention to the hazards of survival at sea following a natural disaster. In addition, her husband, who had been swept out to sea with her, was never seen again.

JMB Smith reported that a woman clinging to a piece of driftwood was rescued 80 kilometres out to sea six days after Hurricane Mitch struck Honduras in 1998.[25] This too, sounds compelling, but this person was also taken out to sea with her husband, and with her three children, and only the woman survived.

It is worth noting that each of these rescues was of one individual only, and in each case the survivors were rescued *at sea* between five and eight days after the event. No-one made landfall on an island or other landmass. Such anecdotal stories, used by researchers to bolster their hypotheses of the potential for accidental colonisation of islands by hominins, in fact indicate the opposite: how *unlikely* it is that individuals, much less groups of individuals, who have been carried out to sea will make landfall.

The Indonesian and Honduras women were very lucky to have been rescued, but what are the chances of surviving at sea if rescue doesn't occur? Firstly, to survive, an individual must have access to fresh drinking water. Humans require fresh water for survival. We can last for weeks without food because the body has many weeks of nutritional reserves, but with no significant ability to store water we can survive only a matter of a few days.[26]

H. floresiensis had an adult stature of only 106 centimetres,[27] equivalent to a modern four-to-six-year-old child. A child of that age needs at least 1.7 litres of water a day to survive.[28] And presumably a greater amount of water would be needed in the high heat and humidity that *H. floresiensis* individuals would have encountered in their tropical regions, unless they had a very different metabolism to us. Now, rainfall would have been plentiful in the monsoon season, occurring daily. Had hominins been swept off land on a natural raft of vegetation, the availability of fresh water is unlikely to have been an issue. That said, the rainwater still had to be collected somehow, and as far as we know, *H. floresiensis* did not make any containers or other vessels. Even if they did, how likely would it have been that these were on the vegetative mat when it was dislodged by a tsunami? So for *H. floresiensis* to get fresh water, we would have to assume that rainwater pooled on the vegetative mat and could be scooped up by hand, or perhaps collected in a large leaf or something similar. If the hominin was floating on a tree trunk, however, obtaining fresh water from downpours would have been much more challenging.

Beyond the crucial act of staying alive in such conditions, normal functioning is also important. Mild-to-moderate dehydration can affect our level of alertness and consciousness, causing sleepiness, muscle weakness, headaches and dizziness. Severe dehydration causes arm, leg and visual weaknesses, and the body finds it hard to regulate its core temperature.[29] This results in low blood pressure, confusion, rapid heartbeat, fever, delirium and eventually unconsciousness.[30] For *H. floresiensis* to achieve landfall on Flores, it would have been essential to have enough fresh water at hand to both survive *and* to ensure individuals functioned well enough to retain consciousness, at the very least.

Also critical to survival is *not drinking* sea water, which is dangerous and can even lead to death. Sea water contains 3.5 per cent salt, but urine cannot contain more than 2 per cent salt. If we drink sea water, our kidneys are unable to flush out the salt they have filtered from the blood, unless a large amount of fresh water is subsequently consumed. In an untreated person, kidney failure ensues, followed by death. We cannot know if early hominins were aware of this 'hidden' danger, so we must assume that to reach Flores, *H. floresiensis* or its predecessors knew not to drink sea water even if fresh water was unavailable.

Survival is then dependent on making landfall. To arrive on Flores, the vegetative mat on which the species was being transported had to stay intact for the duration of the voyage. It had to be dense and strong enough to withstand buffeting by turbulence—we know seas can get rough during monsoonal storms. The sad fate of individuals should the mat have fallen apart was likely to be drowning or shark attack.

Let us assume for the moment that these ominous events did not occur and the individuals reached an island. When I think of castaways being washed up on a beach, I immediately think of Robinson Crusoe. If memory serves me correctly, he was shipwrecked within wading distance of an island; it all seemed relatively easy. But we cannot assume that such an idyllic landing awaited any hominins arriving on the shore of Flores. For one thing, what kind of coastline would they have encountered? We do not know the answer to this question, but today, some of that shoreline is made up of steep rocky cliffs that plunge deep into the ocean. In those places, unless hominins could somehow self-propel themselves into another bay, the effort to attain dry land would have been hopeless; that is, although hominins might have floated on a vegetative raft to within sight of Flores, making landfall might not have been guaranteed. On the other hand, if sandy, muddy or marshy bays were encountered, getting ashore would have been easier.

Because *H. floresiensis* lived for at least 40 000 years on Flores, and we have evidence establishing that hominins were at Mata Menge from at least a million years ago, we know these populations were viable. By a 'viable population', I mean one that has enough genetic diversity among breeding pairs to ensure a high probability of survival over a relatively long period. To achieve this on an island, the initial colonisation would have to depend on a series of interrelated events. The first is that the incident in question, such as a tsunami, dislodges male and female individuals of reproductive age. Alternatively, if a different mix of individuals, such as all males or all females, was swept away, then establishing a viable population would depend on another tsunami occurring pretty soon afterwards and carrying off the right mix of sexes in order to potentially produce offspring with the earlier group. In other words, the arrival of individuals need not be at exactly the same time, but fertile individuals would need to arrive in relatively close succession. And according to this scenario, the individuals concerned would have to actually come in contact with each other, no matter which part of the 380-kilometre northern Flores coastline they were deposited on.

So the establishment of a long-lasting, viable population on Flores, after voyaging there from Sulawesi by floating on sea currents, would depend on the right mix of individuals being on a section of land that has the potential to be dislodged during a tsunami or storm action, or on several victims clambering onto a natural raft; and on this floating structure being relatively stable and staying afloat rather than disintegrating under the erosional action of water. It would depend on the floating mat not becoming trapped in an eddy. It would depend on survivors obtaining fresh water and not consuming sea water while travelling to Flores, followed by the safe landing of enough individuals to multiply and thrive. While not impossible, this would all seem to defy the odds.

An alternative scenario we can consider is that *H. floresiensis* arrived on Flores via the sweep of islands that arc there from Sumatra. The island of Sumatra emerged between sixteen and eleven million years ago.[31] The blobs of land to the east of Sumatra appeared around ten million years ago, and from five million years ago these started to increase in size—they now form the islands we know as Java, Bali, Lombok, Sumbawa and Flores.[32] Some historical background on the distribution of animal species among these islands will provide some context here.

In 1845, W Earle published an article in the *Journal of the Royal Geographic Society of London* in which he identified two geographical areas in the South-East Asian–Australian region. One he called the Great Australian Bank, which is the shallow sea that connects New Guinea with Australia. The other he named the Great Asiatic Bank, also a shallow oceanic area, averaging about 30 fathoms (55 metres) in depth, that connects the islands of Sumatra, Java, Bali and Borneo with the South-East Asian mainland, and stretches from the eastern side of Borneo to within 80 kilometres of Sulawesi. It includes the Malay Peninsula and the northern coastlines of Sumatra and Java.[33] Today, we call this underwater bank the Sunda Shelf—the greater landmass that would have been exposed during periods of low sea level in earlier times is called Sundaland.

In 1853, while studying the natural history collections at the British Museum in London, the naturalist and geographer Alfred Russel Wallace noticed that the flora and fauna of some regions of the world were not well represented. One of these regions was the islands of South-East Asia, and Wallace applied for funding to mount an expedition to investigate the area. While he specifically wanted to collect items for the museum, he would also support himself by selling specimens to private collectors and naturalists,[34] a normal and respectable livelihood in those days. And so for

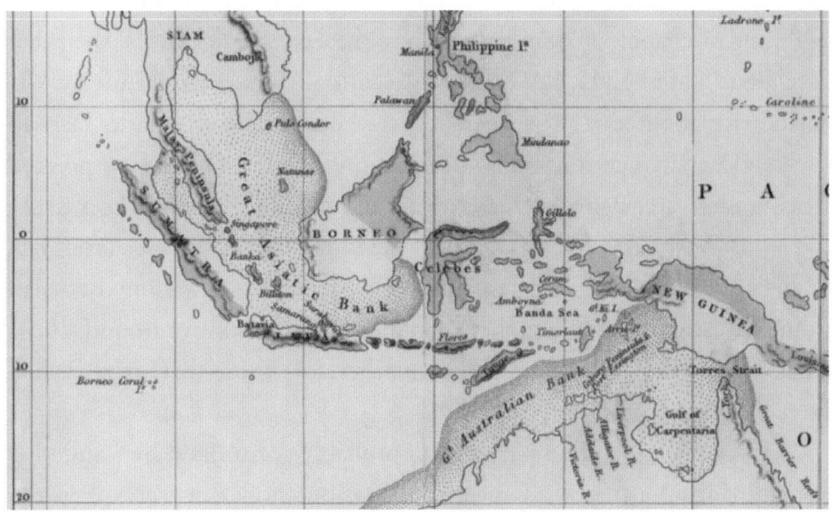

W. Earle's map of 1845 showing the Great Asiatic Bank. Reproduced with the kind permission of the Royal Geographical Society of London.

eight years—from April 1854 to April 1862—Wallace travelled throughout the islands region of South-East Asia and the Malay Peninsula.[35]

Wallace was not merely a collector of rare species, or species previously unknown to the scientific establishment. He was also a keen observer, a scientist of nature. He is famous for developing the theory of evolution, or, as he called it, 'a Law of Natural Descent',[36] at the same time as Charles Darwin—Darwin and Wallace's ideas were presented concurrently to the Linnean Society of London on 1 July 1858.[37] But Wallace was just as absorbed with resolving the question of the odd distribution of species across South-East Asian islands as he was with evolution.

Wallace went on to identify a virtual north–south boundary that separates the faunal types of the Philippines and western Indonesia from the eastern Indonesian islands. He explained his findings in a letter to a friend, Walter Bates, on 4 January1858: 'The boundary line often passes between islands closer than others in the same group. I believe the western part to be a separated portion of continental Asia, the Eastern, the fragmentary prolongations of a former Pacific continent.'[38]

Today, we call this species boundary the Wallace Line (see Figure 1 at the beginning of chapter 1). It closely follows the eastern edge of the shelf that Earle identified in 1845. But Wallace was initially unaware of Earle's work, only hearing about it in 1859 when he received a letter from Charles Darwin in which Darwin mentions Earle's publication.[39] Later, Wallace incorporated Earle's underwater bank into his theory and proposed that changes in sea levels must have occurred on a grand scale in the past. When sea levels fell, the shelf identified by Earle would be exposed, greatly expanding the Asian landmass,[40] and thus enabling the Asian species to freely move around. Those remaining on Sumatra, Java and Bali are therefore those that arrived when the islands were connected by land.[41]

So now we can see how *H. erectus* probably got to Java: by wandering overland during a period of low sea levels. But how can we explain the presence of hominins on Flores if we believe they came via the islands of Java, Bali and Lombok? As Wallace deduced, deep channels separate the islands east of Bali from Lombok. Even when sea levels were low in the past, these islands were never joined—they never formed a single landmass. Animal species from Java did not get further east; they did not cross the Wallace Line and they are not on Flores. It is highly unlikely, then, that hominins walked to Flores.

A possible alternative explanation for how hominins got to Flores is that they found themselves making accidental sea crossings. In that case, at least three crossings would be involved: from Bali to Lombok, Lombok to Sumbawa, and Sumbawa to Flores. More crossings, of course, would be required at periods of high sea levels when Bali was separate from Java and Java was separate from Sumatra.

The straits between these islands are relatively narrow and it is no big deal these days to travel between them in motorised boats. But hominins left to the mercy of currents would be another matter entirely. If Brasseur and colleagues, and also Liu and Ding, were correct when they concluded that monsoons were occurring as early as 2.5 million years ago, then hominins would have faced monsoon-driven currents in the straits. We cannot know the exact form or intensity these would have taken, but I can alert you to the problems that would face anyone being swept off these islands today.

Situated between Bali and Lombok is the 20–40-kilometre wide Lombok Strait. Between July and September, the south-east monsoon causes a south-flowing current running at between 3.5 and 6 knots (1.5–3 metres per second). In all likelihood, anyone carried off Bali into the Lombok Strait would find themselves rushed northwards into the Java Sea or southwards into the Indian Ocean, depending on the monsoon season in which the unhappy event occurred. An extra hazard in the strait is the strong eddies that may be encountered at its northern and southern entrances, and there are turbulent sea conditions at its narrowest part.[42] In each case, without rescue, victims washed off Bali would face poor prospects for survival.

I have never experienced boating in the Lombok Strait, but Wallace did. Here is his description of a voyage that was clearly not for the faint-hearted:

> The beach [at Ampanam, Lombok] of black volcanic sand is very steep, and there is at all times a heavy surf upon it, which during spring tides increases to such an extent that it is sometimes impossible for boats to land, and many serious accidents have occurred. Where we lay anchored, about a quarter of a mile from shore, not the slightest swell was perceptible, but, on approaching nearer, undulations began, and rapidly increased, so as to form rollers which toppled over on to the beach at regular intervals with a noise like thunder. Sometimes the surf increases suddenly during perfect calms, to as great a force and fury as when a gale of wind is blowing, beating to pieces all boats that may not have been hauled sufficiently high upon the beach, and carrying away incautious natives. This violent surf is probably in some way dependent on the swell of the great southern ocean, and the violent currents that flow through the Straits of Lombok. These are so uncertain that vessels preparing to anchor in the bay are sometimes suddenly swept away into the straits, and are not able to get back within a fortnight. What seamen call the 'ripples' are also very violent in the straits, the sea appearing to boil and foam and dance like rapids below a cataract; vessels are swept about helpless, and small ones are occasionally swamped in the finest weather and the clearest skies. I felt considerably relieved when my boxes and myself had passed in safely through the devouring surf.[43]

To the east of Lombok lies Sumbawa, with the two islands separated by the 117-kilometre-wide Alas Strait. The currents in this strait run in a northerly or southerly direction in accordance with the monsoons. Anyone washed off Lombok into the Alas Strait would most likely be transported into the Java Sea or the Indian Ocean, with little possibility of making landfall; similar conditions are encountered in the Sape Strait, which separates Flores from Sumbawa,[44] suggesting that the same potential outcomes face individuals accidentally swept off that island.

The means by which early hominins might have arrived on Flores stirred the imagination of the National Geographic documentary-makers who were filming at Liang Bua in 2004.[45] They thought that Sumbawa was the most likely place from which the early hominins set off for Flores, with bamboo rafts being the most likely means of transportation. To test this theory, they set up an experiment with the Liang Bua team. They would build a raft and paddle it from Sumbawa to Komodo. Robert Bednarik, an experienced rafter and a colleague of Mike Morwood, was engaged to design and organise the construction of a 12-metre bamboo raft, which he did with the help of seventeen Sumba people after sourcing 5 tonnes of bamboo from Bali. Aboard for the trip were Thomas Sutikna, Wahyu Saptomo, Mike Morwood and Bert Roberts. Crewed by a dozen fishermen, it took the raft eleven hours to cross the 22-kilometre-wide channel.

The trip took a toll. Strong south-flowing currents swept the raft south of Komodo, resulting in an exhausting paddle to get back to the leeward side of that island. One of the passengers became violently seasick and the support team had to remove him from the raft. Two other passengers became exhausted after six hours and they, too, were rescued from the situation. Then, as the raft approached the Komodo shoreline, the crew was faced with wild conditions, with waves crashing over rocks, and there was no obvious landing place in sight. This fraught situation was further complicated when Mike tripped and broke his ankle.[46]

Bert Roberts later made an interesting observation about the voyage. The raft, constructed from fresh-cut bamboo, became weighed down and heavy to manage as it absorbed sea water. 'If you are going to build a bamboo raft, it needs to cure for about a year to dry out. Holes in bamboo, the sap channels, seal up when the bamboo dries out,' said Bert, adding that when 'the raft we were on was being returned to Sumbawa, it sank'.[47]

Putting aside the assumption that early hominins could, or did, make rafts, this experiment highlighted the potential dangers. Had the twelve-person crew not been able to turn the raft around and get to the leeward side of Komodo, it is likely they would have been taken out to sea. Had they not been able to manoeuvre away from the hazardous rocky shoreline, their vessel was in danger of being smashed to pieces and its occupants thrown into the ocean. How much more dangerous would this journey have been for hominins entirely at the mercy of currents?

Yet hominins did get to Flores. So let's imagine how this could have happened, ignoring for the moment the danger of the north–south monsoonal currents.

Say, for example, that a number of hominins are on a beach on Java, gathering shellfish. A tsunami races in, or a gale occurs such as the one Wallace described, and the hominins are washed away along with uprooted trees and debris from the coast and the island's hinterland. As luck would have it, all the hominins manage to cling to the debris until, with an extra dash of good luck, they are deposited on a neighbouring island. Years later, the same group endures another tsunami and are lucky, again, to clamber onto some floating debris and, again, are washed up onto the next neighbouring island. Could this misadventure have happened to the same group on four successive occasions, facilitating them crossing four straits, to eventually arrive on Flores? I think not.

A more plausible scenario is that a group comprising males and females of breeding age are swept out to sea following a tsunami and, luckily, they all land on the next island. The descendants of the ever-growing group then populate the island. At some stage, a group of these descendants become the victims of another tsunami, with some arriving on the next island to the east. If both males and females of breeding age are again part of this group, it would grow in numbers and, again, the island would become populated. And so the occupation of each island in turn would happen over the generations until a descendant group of hominins found themselves on Flores. Yet even this scenario has its flaws. How likely is it that the groups washed off the islands included males and females each time? Also, if the survivors could not see land, how could they know that they needed to somehow counteract the north- or south-flowing current they found themselves in? Would they even have had the strength to do so?

Furthermore, we have seen from historical records that tsunamis can occur in close succession or they can take place as many as thirty, forty or more years apart. Again, we do not know the frequency of tsunamis at the time hominins first lived in the region, but if there were thirty-year intervals between these events, it would seem unlikely that castaways would wash up on Flores within one generation such that they could produce offspring.

I have one last thought to share on the vexing question of how *H. floresiensis* arrived on Flores. At the time of the first toolmakers, there were several animal species on Flores, including a giant 1.8-metre-tall bird;[48] stegodonts; a giant tortoise; a small crocodile; a giant rat the size of a small dog; and a large predatory reptile, the komodo dragon, which still lives on Flores.[49] Meijer and colleagues and van den Bergh and colleagues have noted that each of the early animal species on Flores was capable of making sea crossings. The ancestor of the giant stork can fly.[50] Tortoises can float. Stegodonts, if they had the abilities of their modern relation the elephant, could have swum. Komodo dragons, too, can swim. In this regard, the odd species out is *H. floresiensis*.

Here, we return to my friend Julie's speculation: could *H. floresiensis* have swum to Flores? If we want to get some idea of whether the hominins might have been swimmers, the best we can do is look at our closest relatives, the apes, and ourselves. We modern humans are not innate swimmers. Once we're out of our depth in water, we will drown unless we've been taught the basics of swimming, or at least of staying afloat. Nor does it seem that apes are innate swimmers. But evidence of their potential swimming ability comes from observations made of a captive orangutan and chimpanzee studied by the researchers Renata Bender and Nicole Bender (University of Bern).[51]

The orangutan had been regularly exposed to a pool as an enrichment activity at a private zoo; specifically, it had been taught to dive and swim. At the age of seven years, it began to swim underwater from one keeper to another, and it was later able to fully submerge and swim unassisted for a short distance in the pool. The chimpanzee, meanwhile, was brought up in a human environment. He became accustomed to water play and would submerge his head, always covering his eyes and nose—the longest time he spent underwater was fifteen seconds. When fully submerged, he displayed a great variety of play behaviours. To return to the surface, he vigorously kicked with his legs.

It is interesting that the orangutan and the chimp seemed happy exploring a water environment, but, as Renata and Nicole Bender point out, this does not mean that they are natural swimmers. So it is difficult to argue that the ancestors of *H. floresiensis* had an innate ability to swim. In turn, it would seem unlikely that *H. floresiensis* individuals swam to Flores, despite those long feet.

The fact is that none of the scenarios explored here work as a model for the arrival of *H. floresiensis* on Flores. Whether it's travelling between Sulawesi and Flores via the Makassar Strait, or crossing sea barriers between Bali, Lombok, Sumbawa and Flores, the proposed solutions to this mystery remain unconvincing. How, then, can we resolve this?

Perhaps we cannot because we simply do not have the full picture yet. We know that the animals on Flores in *H. floresiensis*' time could have reached the island by sea. If we could work out where each species came from, we could assess if any of those routes might also have facilitated the arrival of *H. floresiensis*.

It was while still immersed in an early draft of this chapter in June 2019 that I attended the inaugural Asia Pacific Conference on Human Evolution, held at Griffith University in Brisbane. There I met Glenn Marshall who, along with his colleagues in The First Mariners team, tests sea current directions using a man-made structure that replicates flotsam. Inspired by the question of the arrival of *H. floresiensis* on Flores, Glenn suggested that we test for different possible routes by which the species may have voyaged by launching the flotsam device from the southern shore of Sulawesi and that of other islands. This project is still in the pipeline. We could supplement this approach by using advanced computer modelling of possible past ocean currents. This method has already been applied to help figure out how Indigenous peoples may have voyaged to Australia.[52]

I recall Mike Morwood occasionally musing about *H. floresiensis* arriving at Flores via an alternative route: floating from the Philippines along the currents that flow to the east of Sulawesi. This would have entailed a series of very long and probably hazardous accidental journeys. But because of the baffling issue confronting us, we need to consider anything we can that might resolve it.

That includes thinking outside the square and contemplating whether hominins may have fashioned rafts or other watercraft with the intention of making sea crossings. Here, Bert Robert's observation may be pertinent. If hominins did make bamboo rafts, it's possible they would have had to plan the voyage a year ahead to allow time for the harvested bamboo to dry out. Of course, the idea that early hominins could think so far ahead to realise the ambition of voyaging is a radical proposal for palaeoanthropologists, including me, to consider. It takes us back to the heart of the enigma: if hominins were making deliberate, pre-planned journeys between islands on seaworthy watercraft a million years ago, why would *H. floresiensis* have become isolated? How could the population have become cut off on the island for a sufficient length of time to evolve into an entirely distinct species if it was possible for them simply to sail back to where they came from, or if multiple generations of other seafaring hominins from the same species could also reach Flores?[53]

Thinking outside the square, I cannot get those long flipper-like feet of *H. floresiensis* out of my mind. Now, if only we could come up with some way of testing if the species *could* swim ...

6

Big surprise in the Philippines

Homo luzonensis

In 2019, what began as a few random finds around the island of Luzon eventually turned the Philippines into a human-evolution sensation. What's happening there draws a parallel with the discovery of *H. floresiensis* in Indonesia. Could island South-East Asia be the new evolutionary frontier?

The Filipino connection began a world away in 1888, when a German geologist, Edmund Naumann, noticed a stegodont tooth among the collections in the Anthropological-Ethnographic Museum in Dresden. The tooth was labelled as having originated from the island of Mindanao. Naumann placed the fossil within a new species, *Stegodon mindanensis*.[1] This was the first indication that the stegodont, an ancient, now-extinct mammal, had made it to the Philippines. Naumann's observations, however, went largely unnoticed, languishing in an obscure publication.

The tooth turned out not to be an isolated find. During the early 1900s, keen-eyed workmen, collectors and geologists occasionally brought the odd fossil tooth or piece of bone into the Bureau of Science in Manila. In fact, right up until the 1930s, random fossil discoveries around Luzon quietly found their way to the bureau to be stored. These included a fragmentary molar of an elephant, a large piece of stegodont bone, a deer horn, thick bone from a large prehistoric turtle, fragments of ivory from a stegodont tusk, and from the mountains of Cagayan, sections of a jawbone (including molars) from a rhinoceros.[2]

Many of these fossils, once duly labelled and put away, lay forgotten. That is, until 1935, when leading paleoanthropologists Professor Henry Otley Beyer and Professor Ralph von Koenigswald, while attending the second Far Eastern Prehistory Conference in Manila, took the opportunity to delve into the collection at the Bureau of Science. For these professors, as for all researchers, it was an exciting prospect to rediscover fossils in the depths of museums. But more than that, they recognised the importance of what they were doing, and they soon started publishing their findings. By the early 1950s, it had become clear that, as Beyer wrote, extinct rhinoceros, elephants and stegodonts had roamed the Philippines during the Middle Pleistocene 'not less than 250,000-300,000 years ago'.[3]

Professor Beyer remained intrigued by the fossils from the bureau and in 1957 invited Professor von Koenigswald back to the Philippines to have a look at the site in Cagayan where the rhinoceros molars had been found. Von Koenigswald undertook the trip with Larry Wilson (who had found additional fossils there), a Mr Shantz (an archaeology student) and DG Kelly (California Institute of Sciences). They found 'four or five stone implements' from a Pleistocene level but no Pleistocene fossil bones. Wilson had also discovered in the same place some objects called tektites. This was the second time that tektites had been found in association with fossil bones of stegodont and elephants, the first being the *H. erectus* site of Trinil on Java.[4]

Tektites are small pieces of natural glassy objects. They form when a large asteroid or meteorite hits the Earth at such speed that the explosive impact releases enough energy to eject melted soils and rocks out through the Earth's atmosphere. The melted material cools and solidifies, and the glass-like objects plummet, showering hundreds and thousands of kilometres of terrain.[5] The combination of tektites with fossil stegodont and elephant bones at the Cagayan site indicated to Von Koenigswald and Beyer that the Trinil and Cagayan sites were contemporaneous. The stone tools at the Cagayan site, then, were thought to be of Middle Pleistocene age.[6]

Through the 1950s and 1960s, archaeological exploration and large-scale excavations proceeded apace in the Philippines, with the country's National Museum (the new incarnation of the Bureau of Science) taking a leading role under the head of the Anthropology Division, Professor Robert Fox. But it was not for several decades that things really took off in terms of human evolution research. Two key discoveries were eventually made, the first of which concerned a long-dead rhino that left the scientific community reeling.

In 1999, Professor John de Vos (Natural History Museum, Leiden) decided to accept a longstanding invitation from Angel Bautista (Senior Researcher at the Archaeology Division, National Museum University of the Philippines)[7] to visit the Philippines. De Vos recalls:

> I first visited the sites from which we know that there are fossils. The most promising site was Cagayan Valley, but I didn't have money to excavate there. There I found artefacts, bones of stegodont, elephant and tektites.[8]

De Vos couldn't forget those bones, but it would be fourteen years before he could return:

> Later I had a French PhD student, Thomas Ingicco, a zooarchaeologist, who was on an exchange research programme [in the Archaeological Studies Program at the University of the Philippines]. He invited me to the Philippines in 2013 to give some lectures. After delivering my talks I had some time and some money to go with Thomas to Cagayan Valley. There we found a species of the same genus *Celebochoerus* [an extinct form of wild boar] as in Sulawesi. In 2014 I arranged some money by inviting Gert van den Bergh, my former PhD student, George Lyras from Athens, a student of mine, and of course Thomas. They brought enough money in and we could excavate. At about 2m deep we found a partial skeleton from a rhino with butchery marks, artefacts and tektites. It was published in *Nature*.[9]

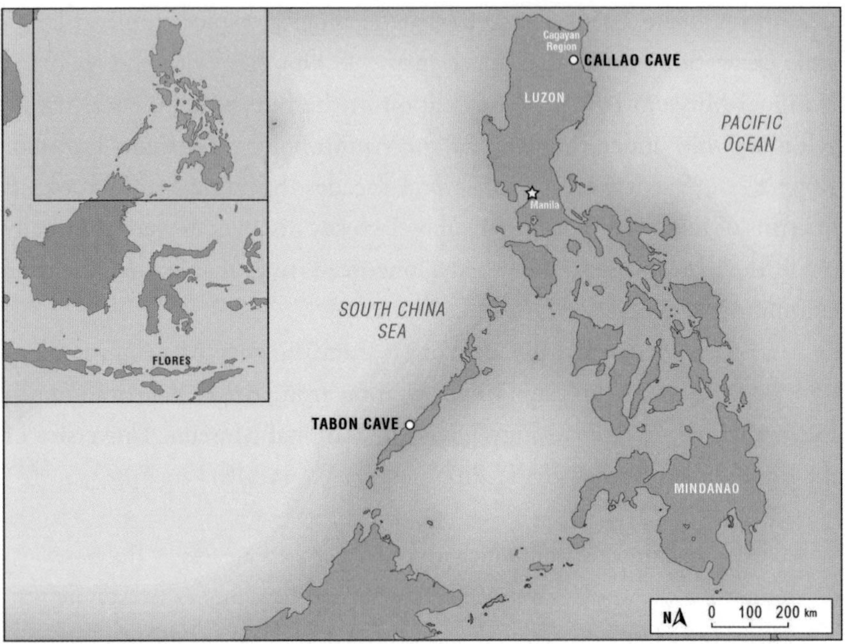

Map of the Philippines showing the location of Callao Cave and Tabon Cave. Prepared by Geraldine Cave.

It was the butchery marks, found on some ribs and legs, that made these bones so special. Someone, or a group of people, had deliberately cut those bones, probably to strip off some flesh and eat it. Someone had also smashed the rhino's two front-leg bones, perhaps to access the highly nutritious marrow inside. The culprit was no modern human, though, because the bone is dated to 709 000 ± 68 000 years ago.[10] Who were these rhino-eating hominins?

Also on Luzon, the Filipino archaeologist Armand (Mandy) Mijares (National Museum of the Philippines), who had been excavating caves for many years, became particularly interested in the period of human history when, around 10 000—12 000 years ago, people in several parts of the world started switching from hunting and gathering to farming. He wanted to know the 'how', 'when' and 'why' of this transition in his own country, enormous questions requiring a lot of research, which saw Mandy start his PhD in 2003 at ANU under Professor Peter Bellwood.

With a lot of archaeological knowledge and experience behind him, Mandy's choice of location to dig was Callao Cave, about 300 kilometres north of Manila. It was one of forty-three caves identified by the National Museum of the Philippines as having archaeological potential during extensive surveys in 1976–77.[11] The cave is impressively large, the biggest in the area, and it is quite a tourist attraction. It has ceiling heights ranging from 10–45 metres and is so spacious that a chapel was built in the second chamber in the 1970s. Archaeologist Maharlika Cuevas and a team from the National Museum had partially excavated it in 1979 and 1980.[12]

Mandy's team headed to the cave in 2003. But how do you decide where to start digging in such a huge cave? Peter Bellwood had the answer. Pointing to a spot at the base of the east wall of the cave, he said: 'There's a nice flat area over there, next to the cave wall. Good potential for discovering archaeological remains. Let's get stuck in!'[13] And dig they did. At a depth of 1.3 metres they found stone tools, burnt animal bones and a hearth. When these were dated, the results showed they had come from Late Pleistocene strata that was 25 968 ± 374 years old—way before farming emerged anywhere in the world.[14]

To further investigate the promising finds in Callao Cave, a partnership was established between the University of the Philippines Archaeological Studies Program, the National Museum of the Philippines and the Australian National University. At this time there was no intention of exploring the earlier history of the cave.[15] The focus remained firmly on the research issue: the transition to farming in the Philippines.

Mandy happened to be back at ANU in 2004, where he was continuing his analyses, when Mike Morwood turned up to give a public lecture about a new species, *H. floresiensis*, that his team had discovered in Liang Bua cave on Flores. Most of us were astonished by the small, archaic-looking hominin bones Mike talked about. What impressed Mandy, however, was the sheer depth to which the Liang Bua team had excavated: 11 metres. Mandy realised then and there that 'you just dig deeper', and he thought: 'I can do this!'[16]

By 2007, Professor Bellwood had secured an ARC grant that enabled the team to return to Callao Cave to dig deeper. A decade later, Mandy

Callao team, 2007; from left to right: Sandy De Leon, Archie Tauzon, Mandy Mijares, Nida Cuevas, Domeng Pagulayan and Kardo. Image provided by Armand (Mandy) Mijares.

recollected: 'Inspired by the work of Mike Morwood at Liang Bua, Flores, I went back to Callao and excavated my original trenches to a greater depth with the hope of finding evidence of early hominin activity.'[17]

Starting where the 2003 dig had left off, the team soon encountered a high concentration of bones in a thin layer lying 2.7–2.9 metres deep.[18] This was well below the level in which they had earlier discovered the hearth, burnt animal bones and tools. At the end of the fieldwork season, the team duly packed up all the bones they'd found and took them back to the University of the Philippines to be identified. Zooarchaeologist Associate Professor Philip Piper had just joined the Archaeological Studies Program at the university.[19] Mandy set the box of bones from Callao Cave on Phil's desk and asked him to identify the bones therein, all 807 of them.

Where does a zooarchaeologist start? It's not just a matter of picking out each bone, holding it up and figuring out its species. A zooarchaeologist needs to understand what happened to the bones before they were excavated, how they got into the cave, as well as what happened to them afterwards, and anything else that might be of future importance. This proved a challenging task. As Phil explains: 'The bones were all covered or partly covered in hard sediment and they were very broken up because bones in caves get fragmented by trampling, or when transported by water, and when they get buried [by sediment].'[20]

An elaborate series of details was recorded for each bone, including measurements, and they were inspected for cut marks or trampling marks, and for evidence of being water-worn. Most of the 807 bones were from deer, some were pig bones, and there were also two tooth fragments from an extinct water buffalo. But Phil was on the lookout for something special, because when Mandy had handed over the box of bones, he'd given Phil a heads-up: 'There's a strange one in there, might be human.'[21]

And there it was—a possible human foot bone, albeit covered in a thin layer of calcite and broken into two parts. Phil immediately called Mandy: 'Hey mate, you have human remains!'

'What? Really?'

'Yes, you have human remains.'

It was beers all round that evening.[22]

Tiny though this bone was, it was particularly significant because it had to be older than the material from the 2003 excavation that was dated to the Late Pleistocene. 'We knew we were onto something,' recalls Phil.

The team did a bit of wishful thinking, guesstimating the bone's age as possibly at 50 000—60 000 years old. At that time, the earliest date for modern humans in the Philippines was 48 000 ± 10 000 years old, derived from a human tibia (lower leg bone) excavated at Tabon Cave.[23] The team's initial hope was that their bone would date to earlier than this.

Phil and Mandy now called in Dr Florent Détroit (Musée de l'Homme, Paris), an archaeologist and bone specialist who had long been involved in Philippines archaeology. Working with Florent were master's student Guillaume Champion and, later, Dr Guillaume Daver (Université de Poitiers, Poitiers, France). Florent says:

> I first visited Callao during the 2007 excavation, after our excavation season in Tabon Cave in March 2007, with several European masters students, including Clément Zanolli and Julien Corny at the time, and then I went again some months after to Manila to continue the work on

> Tabon. So, it was after visiting Callao Cave ... that Mandy and Phil came to me with this possible human foot bone that Phil identified among the animal remains found in a deep layer at Callao ... It was indeed a human bone: it was the 3rd metatarsal.[24]

The metatarsals are the long bones of the foot, the ones that join the ankle to the toe bones.

Clément, then a master's student, was there at the time: 'I still remember the puzzled face of Florent when he saw the metatarsal in the aluminium foil it was kept in at that time, and he recognised the bone as belonging to *Homo* but with unusual features for modern humans.'[25]

It was not a straightforward matter to study the bone. Ideally, the layer of carbonate would be physically removed, but as Florent explains:

> Actual cleaning of fossils has always been a very 'dangerous' step in the preparation of the fossils so that they can be studied, measured, etc., especially when they are encrusted by carbonates, because the bone (even fossilized) just beneath is most of the time more fragile than the carbonate.[26]

So instead of physically cleaning bones, researchers these days use sophisticated CT imaging techniques. Florent continues:

> This is really mandatory to be sure to preserve the integrity of all our precious fossils! [Using CT images] you virtually remove this extra material. You obtain a perfectly (though only virtually) cleaned fossil that you can study, analyse and even use to produce 3D printings of the cleaned fossils if you want to handle them and compare them visually with other fossils or casts.[27]

Expecting the bone to be modern, Florent and Guillaume were surprised to find it had some non-modern characteristics. As Florent recalls: 'Many features of this bone do not fit into the variation known for *H. sapiens* (including fossil *H. sapiens*).'[28] For example, the base of the modern human metatarsal is flat or slightly concave, while the Callao Cave bone is convex. The Callao Cave bone is also slightly convex in side view and it has a ridge running along the surface. Apart from its small size, the most striking thing about it is its unusual proportions.[29]

These anomalies were perplexing. From whom did this bone come? To get a handle on this question, Florent and Guillaume compared the metatarsal to the foot bones of primates of medium-to-large body size living in island South-East Asia: macaques, orangutans, gibbons, and *Homo*, including *H. floresiensis.* In these sorts of analyses, bones from the fossil record are also included. For foot bones this is tricky because so few have been discovered. However, there were some for the earliest species in our genus, so a *H. habilis* metatarsal was added into the mix.

The team found that the Callao Cave foot bone was smaller than those of orangutans, and larger than those of gibbons and macaques. Its length was similar to the metatarsals of *Homo*, particularly the small-bodied Philippine Negritos,[30] *H. habilis* and *H. floresiensis*. Its particular form, though, was different from anything previously described in fossil specimens of the genus *Homo* from eastern Africa, southern Europe or the Caucasus. Palaeoanthropologists are wisely reluctant to declare a new species based on a single enigmatic foot bone. The team provisionally attributed the foot bone to a Pleistocene—but probably a small-bodied—human group.[31]

It was increasingly becoming important, then, to directly date the foot bone. It turned out to be 67 000 years old[32]—even older than the team's earlier wishful-thinking guesstimate. Such an early date for humans in the Philippines was quite a find.

The team knew they needed more bones of this mystery group. But when they returned to Callao Cave in 2009 to continue the excavations, they found, well, nothing.[33] Subsequent excavations in 2011 at first did not seem to be going too well either, in that no-one was finding any fossils. Mandy felt despondent until two things happened. First, a foot and a hand bone were recovered from sieving.[34] Sieving is an essential process in which very patient archaeologists sieve all the soil removed during the dig—it sounds tedious, and it would be except for the camaraderie of your fellow sievers. The objective is to find any items that might have been overlooked as the team excavated. This is no reflection on the archaeologists: bones and artefacts can be unrecognisable as they are dug up, especially if, as occurs in Callao Cave, there is a high clay content that adheres to bones and artefacts. The process is that the soil is first dry-sieved through a 4-millimetre mesh. The material that does not go through is then sieved again using water to dissolve the clays and reveal what lies beneath.[35]

The second bit of excitement occurred when Phil Piper saw an upper leg bone sticking out from under a rock. On carefully extracting it, he saw that it was from a juvenile and he recognised it as at least primate-like.[36]

More fossils were to come. Mandy recalls that one lucky student, Archie Tauzon, who had a penchant for listening to the 1960s surf-rock band The Beach Boys as he excavated, discovered a fossil, then another, and another. Something strange began to happen—every time someone in the team played The Beach Boys, they seemed to find another fossil.[37]

One of the fossils was a pedal phalanx, or toe bone. Florent immediately noticed its 'incredibly marked longitudinal curvature'.[38] That is, the toe bone was curved from top to bottom, unlike ours. Some teeth were then discovered. Florent, who was in the excavation square at the time, recalls:

Callao Cave excavation (2011); from left to right: Florent Détroit, Archie Tauzon and Mandy Mijares. Image provided by Armand (Mandy) Mijares.

> When we found the teeth, I immediately noticed their very small size and some striking features such as the 3-rooted premolars. All these features which could be easily observed immediately ... gave me the feeling that the assemblage would most probably not fit into the frame of the species *H sapiens*.[39]

Over just a few days in August 2011, eleven fossils were extracted from the thin, concentrated bone layer.[40] What was being uncovered in the Philippines was starting to shape up very much like the discovery of *H. floresiensis*. And the Callao Cave discoveries could well result in a similar bombshell. But at the time, the finds were a well-kept secret. So it came as a complete surprise when, on 31 August 2012, Professor Peter Bellwood popped into the office of Colin Groves at ANU, where Colin, Bill Jungers and I were discussing our *H. floresiensis* project. He casually said, 'Mandy's here. From the Philippines. He's brought an upper [third] molar from their excavation at Callao Cave that Rainer is going to date. Come and see.'

Mystified but highly curious, we hotfooted it across to the dating lab run by Professor Rainer Grun in the Research School of Earth Sciences. Mandy and Phil had just walked over there. The tooth was tiny—we thought it the smallest adult tooth we'd ever seen, possibly smaller than *H. floresiensis*.[41] This was the 'exciting stuff' Mike Morwood had predicted back in 2005 in the wake of the *H. floresiensis* discovery: 'Island South-East Asia—you've almost got a laboratory situation. Different islands, different environments, hominins could have got there at different times. What a fantastic research prospect! This is exciting stuff.'[42]

But there was more. Later that day, Mandy and Phil showed us images of five other teeth excavated from Callao Cave. In one image, three molars and two premolars were lined up, as they would sit in the palate. Seeing these clinched our initial impressions: we knew that this must be one tiny new hominin. Colin, Bill and I tried to get our heads around what this discovery meant for human evolution, feeling privileged to have been shown the premolar and the images of the other teeth. We could certainly well understand why Mandy and Phil really wanted the tooth dated, bringing it direct from the Philippines for that purpose.

Callao Cave excavation (2015). Image provided by Armand (Mandy) Mijares.

The ANU Research School of Earth Sciences team undertook U-series dating, for which they laser-drilled a row of twelve minute holes into the tooth to extract the datable material. It sounds a bit destructive to drill into a fossil, but at 0.2 millimetres wide and 1–2 millimetres deep, the holes are tiny; they take up only a minuscule bit of space and can hardly be seen by the naked eye. The age of the Callao Cave tooth turned out to be a minimum of 50 000 years. So a tiny hominin lived in the Philippines fairly recently, in human evolutionary terms, and at the same time as modern humans. This really was looking like the *H. floresiensis* scenario.

When the Callao Cave team returned to the site in 2015, they employed their tried-and-true fossil-finding technique. And while playing The Beach Boys, they discovered an upper third molar. They now had three individuals represented among the remains.[43]

The fossil count as of 2020 was seven teeth, two hand bones, two foot bones and the femur shaft.[44] As Phil and Mandy had shown Colin, Bill and me back in 2012, five of the teeth were from one individual. It was possible to know this because each fit neatly beside the others, so these teeth would have been inside the same mouth.

The team now had a foot bone and some teeth with primitive features. These were unlikely to be from a modern human, as the researchers had thought was the case with the foot bone back in 2010. This called for a team rethink about the identity of the Callao Cave remains, examining each element separately and then putting it all together.

Mandy first recruited two experts in tooth structure and form, introduced above: Dr Clément Zanolli, now a researcher from the Paris-based French National Centre for Scientific Research and also the Université de Bordeaux, and Dr Julien Corny from the Musée de l'Homme. Clément is adept at interpreting the internal structure of teeth from scans, and Julien's expertise is the external, or outside, morphology of the teeth. Clément observes:

> First we noticed modern-like features of the teeth: they are small and have a simple form. But the two premolars have three roots. This doesn't occur frequently on moderns and even in the archaeological record they are rare, but they are often observed in fossil hominins. So this was our first clue that the teeth are more primitive than modern humans.[45]

So far, so intriguing, but it was time for more-comprehensive investigations. Zanolli and Corny's first approach was to compare the teeth with those of *H. erectus*. Recapping our evolution history, *H. erectus* is known from Java, Indonesia and, as we saw in chapters 1 and 3, is considered by some to be the ancestor of *H. floresiensis*. Could it also be the ancestor of the Callao Cave hominin? The two researchers noted that a group called 'late *H. erectus*', which might have survived on Java until just 143 000 years ago,[46] had third premolar teeth with the same three-rooted pattern as the Luzon teeth. But *H. erectus* teeth are very large. Zanolli and Corny's thinking was that the Luzon hominin could possibly be a dwarf form of late *H. erectus*, much as had been proposed by some to explain the origin of *H. floresiensis*.

Another idea they considered was whether it could be a Denisovan, which was identified by DNA from a finger bone excavated in Denisova Cave, Siberia (see Appendix A). The Denisovan teeth are larger than those of *H. erectus* and thus much larger than the Callao Cave teeth. The Denisovan molars also have a more complex form.[47] Clément: 'We don't see anything of the Callao Cave teeth in the Denisovans, so we discarded this idea.'[48]

Zanolli and Corny now broadened their comparisons to include the teeth of *A. afarensis*, *A. africanus*, *H. erectus*, *H. floresiensis*, *H. naledi*, *Paranthropus*, early *Homo*, the Neanderthals and modern humans. Professor Yousuke Kaifu (National Museum of Science, Japan, and the University of Tokyo) kindly provided the team with scans of one of the *H. floresiensis* premolars to include in their analyses. The French researchers have found that the shape of the Callao Cave upper molars as a whole is unique. Not only are they small, but they are narrow from front to back, unlike anything in our genus, *Homo*, the australopithecines or *Paranthropus*. Yet some internal structures, and the form of the lumps and bumps on their crowns, resemble modern human teeth.

The premolars are large relative to the molars, more so than seen in *H. floresiensis* and other hominins, except *Paranthropus*. They also have some primitive features, some of which are seen on *H. floresiensis* and some of which are not. The premolars have multiple roots, spread widely as found in *Australopithecus*, *Paranthropus* and the earliest species in our genus, *Homo*, and *H. erectus*. One of the premolars, thought to be an upper fourth premolar, has three roots, which is rarely found in fossil hominins and *H. sapiens*.

If this sounds complicated, spare a thought for the researchers trying to figure it all out. Overall, the odd size of the premolars compared to the molars and the smallness of the Callao Cave teeth is a pattern not seen in any other species of *Homo*.[49] Clément sums it up: 'The molars of *H. floresiensis* and *H. luzonensis* are modern-like but their premolars having three roots—that's old.'[50]

Bones from feet are also useful when looking for similarities and differences between species. But one of the three foot bones excavated in Callao Cave is proving perplexing; 'bizarre' is how Florent Détroit describes it.[51] It is very curved when you look at it side-on, nothing like what we see in the Neanderthals, *H. naledi* or *H. floresiensis*, and it does not have the 'hourglass' form of our toe bones. Phil Piper notes that it 'has a remarkable resemblance to 3.2 million-year-old *Australopithecus afarensis*—really intriguing!'[52]

Another toe bone discovered in the excavations is an intermediate pedal phalanx, which is the toe bone that joins with the tip-most of the toe bones. Unfortunately, these toe bones are not particularly useful in species identification because they are variable, even among *H. sapiens*.[53]

The remaining foot bone is the one that first alerted the team to the antiquity of humans in the Philippines. It is the third metatarsal that the team published on in 2010, provisionally placing it in *H. sapiens*.[54] Since then, however, the third metatarsal of *Australopithecus sediba* has been described. The Callao Cave bone shares some similarities with this two-million-year-old extinct hominin from Africa.[55]

The Callao Cave excavation yielded two finger bones. One is a middle finger bone, technically known as an 'intermediate phalanx', the bone that fits between the first- and second-finger knuckles. It is from the left hand of an individual. It is flat as well as incredibly long for a supposedly small bodied hominin,[56] and it has characteristics seen in the australopithecines and to a lesser extent *H. habilis*, but not in other *Homo*.[57] It is relatively longer than that of *H. sapiens*, *A. afarensis*, the Nariokotome Boy (also known as Turkana Boy),[58] and *H. floresiensis*.[59]

The other finger bone is a distal phalanx, one of the tip-most finger bones. The shape of this bone is within the range of *H. sapiens* and the australopithecines but it's unlike the finger bones of *H. floresiensis* and the Neanderthals.[60] The Callao Cave fossils are proving to be a quite a complicated puzzle.

The upper leg bone that was found protruding from under a rock in Callao Cave is frustratingly incomplete. With the upper and lower parts missing, all that can be gleaned about this bone from CT scans is that it was the bone of a growing individual. The plan now is to get a better handle on the internal characteristics of this bone. As Clément explains:

> The femur is very interesting because even if it is an infant it still probably retains information internally. So from the bone thickness, for example, we can infer its locomotion, if it walked on all fours or on two legs, and the way it was walking. We can also use the foot and hand bones in this way. This is just the start of our work.[61]

Early on, the fact that the enigmatic Denisovans were identified from DNA drawn from a single finger bone prompted the hope that DNA might be similarly preserved in the Callao Cave bones and teeth. Attempts at extracting DNA from another hominin species in the tropics, *H. floresiensis*, had had disappointing results; none was preserved. Nevertheless, the *H. luzonensis* team figured it was worth a try, and Mandy and Florent took the upper third molar to the Max Planck Institute for Evolutionary Anthropology in Leipzig in 2013. Very limited DNA material was recovered from this tooth, however. The Max Planck Institute wanted other hominin remains from the excavations, but Mandy considered the technique too destructive given the small number of hominin remains they had recovered from the cave.[62] So he came up with an alternative: to have DNA tests performed not on the precious hominin bones but on the animal remains found in the bone layers. 'So I first sent cervid teeth for them to test if they can extract DNA,' he says. 'Unfortunately this also failed.'[63]

It is therefore not worth testing for DNA in the *H. luzonensis* hominin remains, at least not in the conventional way. But there has been a remarkable recent development in DNA research. Dr Viviane Slon and colleagues have been able to isolate DNA from, of all things, cave sediment.[64] This they did for caves in France, Belgium, Croatia, Spain and Russia, including the famous Denisova Cave. They have shown that cave sediments contain lots of DNA, including traces of hominins. For example, Neanderthal DNA has been identified in caves that also contained Neanderthal bones, but it has even been found in caves with no signs of Neanderthal visitation. DNA from animals such as hyenas, bovid, deer, horses and related animals that visited the various caves has also been found.

Mandy is inspired by the fact that DNA can be retrieved from cave sediments. In time, he and the team plan to utilise this discovery and collect soil samples from Callao Cave for DNA testing.[65] What a sublime breakthrough it would be if DNA in soil is preserved under tropical conditions. Could we eventually get *H. floresiensis* DNA in this way?

But back to the Philippines. The particular mix of characteristics observed in the Callao Cave hominin remains has not been seen in any other species: this suite of teeth and bones is unique to this group. Under the International Code of Zoological Nomenclature, the researchers involved declared this group of fossils a new species: *H. luzonensis*, named for the island on which they were found.[66] This is exciting, but even more it demonstrates that there were two small, archaic hominin species living in island South-East Asia at about the same time.

In terms of current thinking around human evolution, the Luzon connection has other implications. We see characteristics in *H. luzonensis* that were present in species that existed more than two million years ago in Africa, just as we saw for *H. floresiensis*. Yet both *H. luzonensis* and *H. floresiensis* are known from relatively recently (geologically speaking) times, and both are half a world away from Africa. We have been able to hypothesise where *H. floresiensis* fits on the human evolutionary tree because so much of its skeletal form is available to us. However, although most of the *H. luzonensis* fossil material is clearly diagnostic, and there is ample evidence for it being a new species of *Homo*, there are not enough body parts to make a call about where it fits on the tree or who its ancestors might have been. To resolve this, we need a few more clues from a skull or jaw or two; some leg and arm bones and a shoulder bone would help, too. Fortunately, archaeologists and their colleagues are a passionate, optimistic and patient lot. It looks like the Callao team will be well funded and able to return to the cave to excavate for many fieldwork seasons to come.

We might not know where *H. luzonensis* fits on the family tree, but we do know something about its behaviour. When studying those 807 bones from the excavation in Callao Cave, and knowing that no stone tools had been discovered in the layers that bore the hominin bones, Phil Piper noticed something unexpected. 'We've got bones with cut marks. Not many of them, but what we have is really lovely,' he explains. 'I say lovely because the marks are very clear. There is no ambiguity about them.'[67]

Eleven of the deer and pig bones had cut marks on them. To investigate these, the bones were placed under a scanning electron microscope (SEM). This tool, which can magnify an object from around ten times to up to 30 000 times, can create images of otherwise invisible worlds regarding anything from fungi to stones and bones. And the images produced look three-dimensional, which is one of the beauties of this process. Phil could see a distinctive 'V'-shaped incision that is typical of human-made cut marks. Also visible were 'shoulder effects' along the outside edges of these cut marks, a typical consequence of someone using stone tools to deliberately cut bone. There were also incisions running parallel to the main incision, caused by imperfections in the cutting edge of the blade.[68]

This means that these were the remains of hunted or scavenged animals. So despite not having many hominin fossils, Phil has found out that the bones from the excavations are at least partially left over from hominin activity, and importantly, that *H. luzonensis* was a tool-using hominin.

Many puzzles remain. Where are the tools that produced the cut marks on the bones? The Callao Cave team suspects that the answer to this might well be related to the nature of the archaeological record. The configuration of the bones into 'alignments' within flow channels in the cave suggests that they were transported further into the cave by wet-season water action from what is now the cave entrance area. What might have happened is that the stone implements, being heavier, remained closer to where they were discarded. This is a hypothesis at the moment. It could be that the hominins used organic implements, but some details of the cut marks suggest stone was used.[69]

With the fossil remains of this new species being found in a cave, it would be easy to assume that these hominins were living right there. After all, Callao Cave is huge, with an inviting massive entrance, and offering safe and relatively salubrious accommodation. Mandy has been thinking about the form of the cave, though. For a long time, a large slab of rock on the floor of the cave near the entrance was thought to have fallen from the ceiling. But Mandy has concluded that it was part of a wall that collapsed during the mid-Holocene (a period that extends from roughly 7000 to 5000 years ago), opening up a walk-in entrance well after the bones of *H. luzonensis* had accumulated in the cave.

With this possibility in mind, Mandy suspects that before the wall collapsed the cave was a sinkhole: 'I don't think *H. luzonensis* was living there—I think the bones were washed in. There was no entrance before the mid-Holocene. Even the animal bones have been eroded, evidence that they were washed in.'[70]

During the next fieldwork season, the team will try to find out from where the slab fell. Now there's a thought ... I wonder what they'll find *under* the slab?

And we can't forget about those rhino bones with 'butchery marks' discovered by Thomas Ingicco and colleagues. Who were those rhino-eaters? While it is tempting to think they were *H. luzonensis*' forebears, we simply cannot assume this. They are, after all, 709 000 years old, around 659 000 years older than *H. luzonensis*. At the moment, we have no evidence for any hominins in the Philippines during the 659 000 years between the rhino hunters and *H. luzonensis*.

H. luzonensis certainly presents us with a fantastic research project with a very promising future. We know that with the generous support of the University of the Philippines and the country's Commission for Higher Education, a substantially larger team will return to Luzon to further excavate Callao Cave and others in the region. And it is exciting to know, as I understand Mandy has confirmed, that our Beach Boys–playing, fossil-finding Archie Tauzon will be there when they do. Idly thinking, I have planned the playlist for the best results: 'Wouldn't It Be Nice?', 'I'm Waiting for the Day' and 'It's Just a Matter of Time'. And when more fossils are found: 'Celebrate the News'.

Reactions to *Homo luzonensis*

Despite being on many digs, I have never found a fossil. This is always a thrill for anyone in the world of archaeology, so I live in hope. In the meantime, one thing I've always been curious about is what happens behind the scenes when archaeologists find fossil bones. Do they show people? Do they ask other experts for their opinions? Or do they keep the discovery a secret until they publish on it?

I took the opportunity to ask Dr Florent Détroit these questions when I met him in Brisbane in June 2019 at the Asia-Pacific Conference on Human Evolution. Florent first responded that, yes, his team did show the *H. luzonensis* fossils to colleagues:

> We had various reactions about the fossils, but I can say that all of them [the colleagues] were quite puzzled after seeing them. Some colleagues were very doubtful, telling me that they were probably from a large primate, but not human. But since we know of no other primate than *H. sapiens*—and macaques, most probably introduced by humans—in the Philippines, they would have belonged to an unknown species of large primate that was morphologically very close to a hominin, and able to cross the sea gap and settle on Luzon island ... which does not sound [like] a very parsimonious hypothesis.[1]

Others thought that the fossils might correspond to 'abnormal' *H. sapiens*, though Florent was quick to add that this was not the same kind of argument as those published during the *H. floresiensis* debate (see chapter 3). Instead, behind one or two people's view that *H. luzonensis* might be *H. sapiens* was the notion of widening the definition of *H. sapiens* to include Neanderthals and other recent hominin species.

On the other hand, Florent commented: 'Some were very enthusiastic, seeing a clear parallel with *H. floresiensis* from Flores Island.' Still others wanted to see a skull before they would accept that the fossils represented a new species, to which Florent's response was: 'Well, I clearly do not agree with those palaeoanthropologists who think that only skulls are good and interesting for human evolution.'[2]

In chapter 2 we saw the uproar when *H. floresiensis* was announced. Mandy and the team expected a similar reaction over *H. luzonensis*, but this hadn't happened. 'We had some concerns when we published because of the controversy over *H. floresiensis*, but so far, nothing,' he said, adding with a smile: 'We're hoping to get at least some controversy!'[3]

I can see his point. It is well understood that science thrives on robust discussions, and human evolution is no exception to this. Perhaps *H. floresiensis* paved the way for acceptance of the late survival of hominin species in island South-East Asia. But news of *H. luzonensis* was published only recently, and rarely do announcements of new hominin species go unchallenged. Maybe, then, it is a case of 'Watch this space'. The team might yet have to face the controversy it expected.

Let's go digging

Although we scientists are not naïve enough to think we know everything there is to know about human evolution, the discovery of *Homo floresiensis* has presented an enormous challenge to our predominant ideas. We have had to rethink much of what we thought we knew.

This 1-metre-high, small-brained, long-armed, long-footed species lived in the Indonesian region until at least 52 500 years ago, yet it resembled something we'd have expected to find in Africa from around two million years ago. That it lived so recently implies that we once shared our planet with a very primitive form of hominin. We have long accepted that we once shared our world with Neanderthals, who were similar to us. But *H. floresiensis* living at the same time as us? That is something we really have had to come to terms with.

Evidence for such an archaic species implies a much earlier exit of *Homo* from Africa than our previous assumed view of human evolution encompassed. That is, the strongly supported Out of Africa 1 model has been forcefully challenged by the discovery of *H. floresiensis*. In addition, it seems apparent that *H. floresiensis* could make sea crossings, which is a first in human evolution.

These 'paradigm-busters' have led to some resistance to the acceptance of *H. floresiensis*, with claims that the bones, or some of them, represent diseased modern humans. But none of these assertions has withstood testing. It remains clear that *H. floresiensis* is a new species.

Fast-forward fifteen years from the discovery of *H. floresiensis*, to when another small hominin species emerged in excavations on the island of Luzon in the Philippines and was dated to 67 000 years ago. So far, the discovery of *H. luzonensis* has attracted none of the controversy that surrounded *H. floresiensis*. This is significant. It might suggest that human evolutionists are at last accepting the remarkable implications that arose from the discovery of *H. floresiensis*: that diminutive, archaic species persisted into the time of *Homo sapiens*, and they survived sea crossings in numbers great enough to establish viable populations on islands. These crossings occurred at a very early time, as evidenced by the one-million-year-old stone tools discovered at Wolo Sege, Flores and the 710 000-year-old rhino bones with cut marks recovered on Luzon.

The remains of *H. luzonensis* are too fragmentary for us to make a call as to its ancestry. *H. floresiensis*' lineage is also perplexing. Two scenarios have been rigorously tested: that *H. floresiensis* descended from an early unknown species of *Homo* that lived in Africa, or that it descended from an unknown group of *H. erectus* that arrived on Flores at an unknown time and subsequently dwarfed in response to the 'island rule'.

The most compelling evidence favours the case that *H. floresiensis* is closely related to one of the earliest of our genus, *H. habilis*, sharing a common ancestor with that species. We can theorise that the *H. floresiensis* lineage branched from our family tree at a very early stage in the evolution of *Homo*, possibly as far back as two million years ago, and made it all the way to the Indonesian region, where it remained until recently, outliving its closest relative, *H. habilis*, by a million or so years.

Tantalisingly, we now have the remains of some very small hominins that were discovered in the So'a Basin excavations on Flores, not too far from Liang Bua cave where *H. floresiensis* was discovered. Dated to between 650 000 and 800 000 years ago, the So'a Basin hominins are much older than *H. floresiensis* and, given their small size and the location of the finds, it is logical to think that they could represent *H. floresiensis*' direct ancestors. The fragmentary nature of these fossils, however, means that we must await further finds of this little group before we can figure out if it was part of the *H. floresiensis* lineage.

The alternative idea, that *H. floresiensis* evolved from an unknown group of isolated *H. erectus* on Flores, is not strongly supported. This scenario requires a significant decrease in *H. floresiensis*' body and brain size, and it asks us to accept many other evolutionary reversals to explain its *Australopithecus* and early *Homo* characteristics. No such assumptions are required under the *H. floresiensis*-as-an-early-hominin proposal.

Valid questions remain, though. If *H. floresiensis* evolved from an early hominin group in Africa, where are that group's remains? Should we by now have found bones of the alleged lineage somewhere between Africa and Indonesia? We can tackle these puzzles archaeologically. My first choice of site for investigation would be Java. We have obtained *H. erectus* remains from this island, so there is a fairly good chance that other hominin remnants have also been preserved there. My second choice would be the Zarqa Valley in Jordan, where those two-and-a-half-million-year-old stone tools were discovered. Someone made those tools, and their fossilised remains might well be lying in wait for an enthusiastic archaeological team to unearth.

I would also focus on the African landmass. But where to start? The various 'cradles of human evolution' in East Africa and South Africa have proved to be rich sources of hominin fossils that have contributed so much to our existing knowledge. Two new species, *A. sediba* and *H. naledi*, were recently discovered there. Perhaps the remains of *H. floresiensis*' ancestors are also lying there, waiting to be revealed. It would be well worth paying close attention to Chad and Morocco, too, where significant hominin fossils have also been found.

The lands between Africa and South-East Asia are vast. Tackling them archaeologically would be a formidable task. So in the first instance, I would invite geologists onto a team to identify the regions or areas likely to encompass one-to-three-million-year-old landscapes, then formulate a strategy for archaeological investigations in those locations.

One additional idea: *H. floresiensis* and *H. luzonensis* were successful island-colonisers, so perhaps it is time to concentrate on any islands between

Africa and South East-Asia that lie close to a coastline, including that of the Mediterranean.

We have seen outstanding advances in the use of DNA in elucidating the ancestry of hominin species, such as the Denisovans and Neanderthals. Unfortunately, DNA has not survived in the *H. floresiensis* bones or in the fossil animal bones from the *H. luzonensis* excavations. This hereditary material simply does not survive for long in skeletal remains in the tropics. We had thought that was the end of it, that we simply could not resolve the ancestry of these species using this technique. But research in this area is racing ahead, and DNA can now be identified in soil (as mentioned in chapter 6)—although, again, it is unlikely to survive in the earth of the tropics.

More promising is the fact that DNA has recently been identified in stalagmites.[1] It appears that if an animal or human even just brushes against a stalagmite, their DNA can rub off and become sealed in the deposit relatively quickly. What a sublime opportunity it would be to test whether DNA survives in stalagmites in caves on Flores. This possibility has not been lost on me and my Indonesian and Australian colleagues. We have initiated a pilot project with the Max Planck Institute to see if any DNA has been preserved in samples of stalagmites that are available from an ongoing climate history project on Flores. At the time of writing, the global pandemic had intervened in this effort, but I look forward to the time when we can resume our investigations.

It has been such a privilege to work on *Homo floresiensis*. I am excited about the future prospects for research, and I eagerly await the conquest of COVID-19 so that researchers everywhere, and the world at large, can get back to work and a normal life.

Appendix A

Cousins by the dozens

Way, way back

We know that there once was an ancestor of both the human and chimpanzee lineages. This forebear likely lived before about 7–8 million years ago when, according to genetic evidence, the human and chimpanzee lineages split.[1]

Although we don't know exactly who this ancestor was, the discovery of fossil teeth and a jaw near Athens in 2017 has given us some tantalising clues.[2] These fossils have been dated to 7.37–7.11 million years ago,[3] around the time of the split. Called *Graecopithecus freyberg*, they might represent the common ancestor of the chimps and us. Why? It's all in the tooth roots. It might seem rather an obscure thing, but the form of the root of the premolar in our jaw seems to be diagnostic for our hominin lineage.[4] While great apes typically have two or three separate and diverging premolar roots, the roots of *Graecopithecus* converge and are partially fused—a feature that is characteristic of modern humans, early humans and several prehumans.[5]

Just who came earliest in our direct lineage is not clear. There are three contenders so far and each is hotly debated. They are considered contenders because they all walked upright on two feet, one of the trademarks of the human lineage. The earliest to have this trait were *Orrorin tugenensis* (six million years old),[6] *Sahelanthropus tchadensis* (6–7 million years old)[7] and *Ardipithecus* (5.6 million years ago).[8] The ages of *Orrorin* and *Sahelanthropus* are very close to that seven-million-year-old timeframe geneticists have identified for the split between the human and chimp lineages. The players in the *H. floresiensis* story evolved much later, about 3.7 million years ago.

A note about the dates of species before we proceed. It is unlikely that the very first individual of a species to evolve, or the last to have existed, will be found in the fossil record. All dates for species, then, may be considered minimum dates—each species could have evolved earlier than the current evidence implies, and each could have survived for longer.

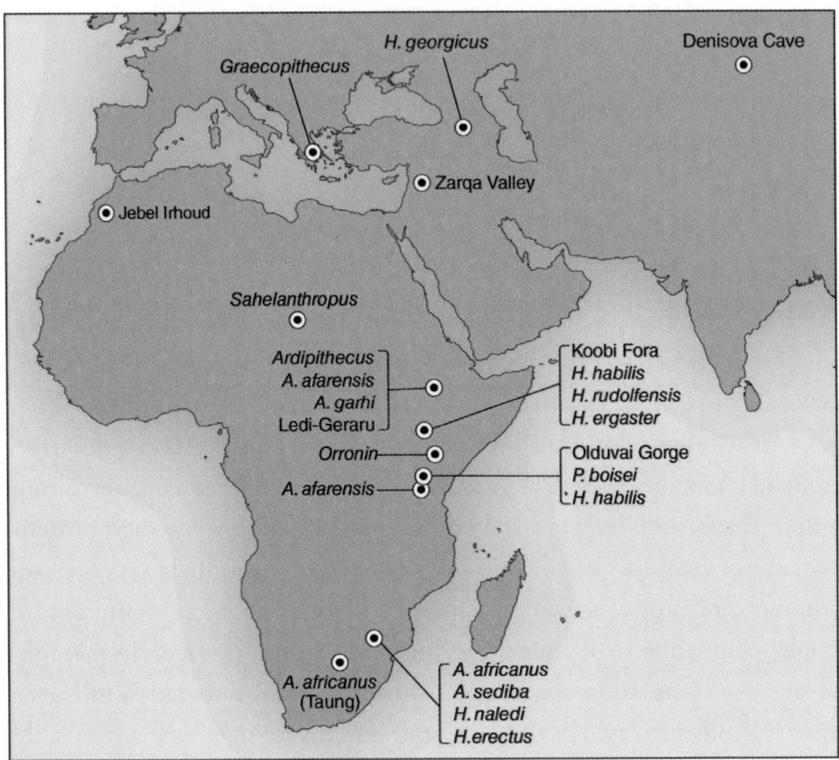

Map showing locations of species and places discussed in this chapter. Prepared by Geraldine Cave

The australopithecines

The australopithecines are the closest relatives of our genus, *Homo*, and comprise a number of species that lived in southern, eastern and central Africa from around 4.7 million years ago to about 1.8 million years ago; they are not known to have subsisted anywhere but Africa. They differ from species of *Homo* in their relatively small, grapefruit-sized brains, in the way the face projects forward and the forehead slopes back, and in the size of their (massive) jaws compared to their skulls. In most species of *Australopithecus*, the arms were relatively long compared to the legs—somewhat chimp-like—and the bodies were barrel-shaped. The males were taller than the females. The australopithecines were upright walkers and might or might not have been tree-climbers.[9]

There are eight species of australopithecines, of which four play a part in our story: *A. afarensis*, *A. africanus*, *A. garhi* and *A. sediba*.[10]

Australopithecus afarensis

The earliest of the australopithecines was *Australopithecus afarensis*, which lived 3–3.7 million years ago in the region that now comprises Ethiopia, Kenya and Tanzania.[11] Archaeologists have found part-remains of more than 300 individuals.

A. afarensis, nicknamed 'Lucy', was an upright-walking hominin with human-like teeth.[12] In terms of its other characteristics, however, it was somewhat ape-like. It had a brain capacity of 428–485 cubic centimetres,[13] which is small compared to our 1100–1500-cubic-centimetre brains. *A. afarensis* had a low forehead, with a bony ridge above the eyes. Its face projected, although it was flattish, with a flat nose. There was no chin; rather, the jaw just sloped back. *A. afarensis* also had very short thighs, powerful arms, and forearms that were long compared to the upper arms (again, somewhat like chimps). The foot and hand bones were curved, and

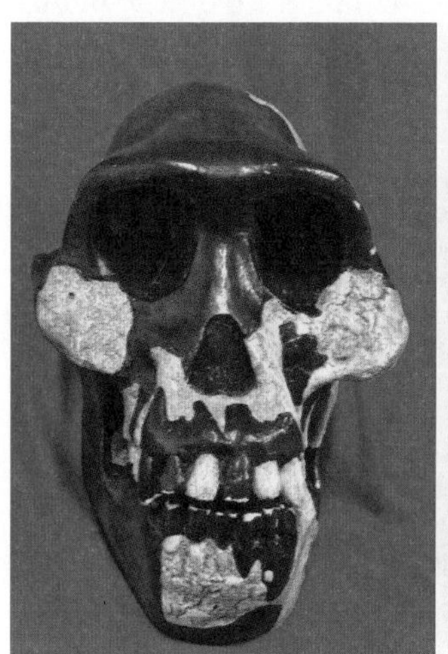

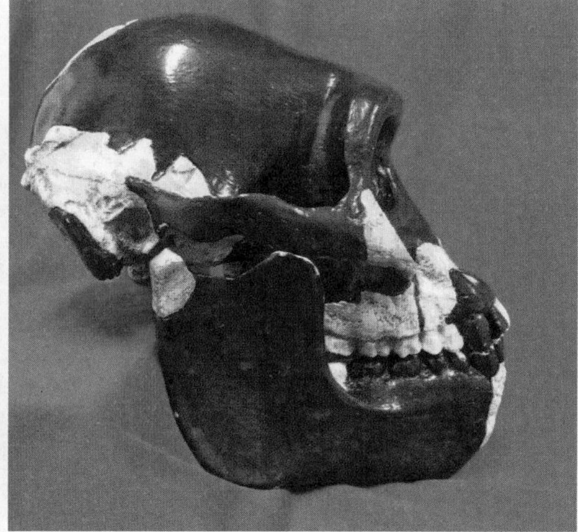

Left: *A. afarensis*, replica model at the School of Archaeology and Anthropology, ANU, front view. The black infill represents filling for missing bone and to hold the original bone fragments together. Image provided by Debbie Argue.

Right: *A. afarensis*, replica model at the School of Archaeology and Anthropology, ANU, side view. The black infill represents filling for missing bone and to hold the original bone fragments together. Image provided by Debbie Argue.

this hominin had long toes—some scientists have inferred from its hands and feet that *A. afarensis* was adept at climbing trees. Standing at about 151 centimetres, the males were taller than the 105-centimetre females. (Also see Appendix B.)

Australopithecus africanus

In 1924, Professor Raymond Dart discovered a small partial skull in rock debris from a limestone quarry near Taung in South Africa. He formally described the skull and named it *Australopithecus africanus*, although we sometimes see it referred to as the 'Taung Child'. It was the first fossil hominin to be discovered in Africa. (Also see Appendix B.)

A. africanus' earliest appearance in the fossil record is three million years ago and its last appearance is around 2.4 million years ago.[14] It lived more than 5000 kilometres to the south of *A. afarensis*' locale.

A. africanus had a combination of human-like and ape-like features. Brain size ranged between 400 and 550 cubic centimetres.[15] Its arms were relatively long in relation to its legs. *A. africanus* walked upright, but its shoulders and arms show that it was also a tree-climber. Like *A. afarensis*, individuals were short: males are estimated to have been around 138 centimetres tall and females 115 centimetres.[16]

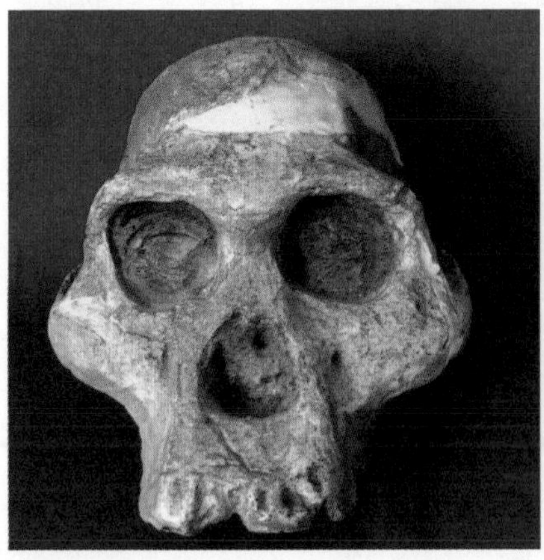

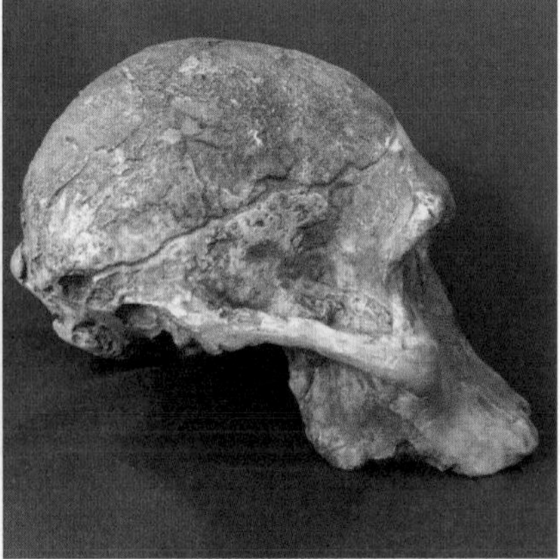

Left: *A. africanus*, STS 5, cast at the School of Archaeology and Anthropology, ANU, front view. Image provided by Debbie Argue.
Right: *A. africanus*, STS 5, cast at the School of Archaeology and Anthropology, ANU, side view. Image provided by Debbie Argue.

This new discovery's tiny brain did not fit with conventional views of human evolution because people thought that brain size would have increased before upright walking evolved, but *A. africanus* showed us that the opposite happened. It took nearly twenty years for the scientific community to accept *A. africanus* as a distinct species.

Australopithecus garhi

The fossil hominin bones discovered in Ethiopia in the 1990s were announced as a new species, *Australopithecus garhi*—the word *garhi* means 'surprise' in the language of the local Afar people.[17] *A. garhi* was declared a new species because the skull, jaw and teeth showed differences from other australopithecines. The other skeletal bones discovered at the time, however, cannot be positively attributed to this species.

Dated to around 2.5 million years ago, *A. garhi* might have been around at the same time as *A. africanus*, but being in different regions—one at the southern tip of the African continent, the other much further north—it is unlikely they ever set eyes on each other.

Australopithecus sediba

Another australopithecine that is part of our story was discovered just a few years ago, in 2013, in South Africa. It is called *Australopithecus sediba*. The word *sediba* means 'fountain' or 'wellbeing' in the deSotho language.[18]

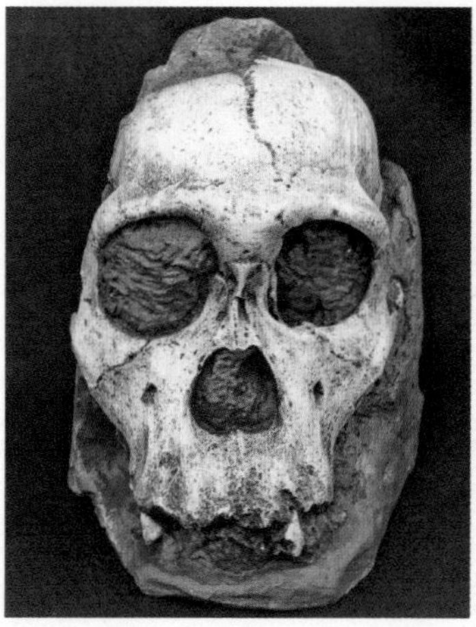

A. sediba, cast at the School of Archaeology and Anthropology, ANU. Skull is within the rock in which it was embedded. Image provided by Debbie Argue.

So far, *A. sediba* is known from only the one site: the Malapa Cave, which is about 30 kilometres from where the Taung Child was discovered. *A. sediba*, then, would have roamed the same area as *A. africanus*, albeit somewhat later, around 1.95–1.78 million years ago.[19]

Two partial skeletons belonging to one adult and one child, as well as some other bones, are known from this species. The brain size of the young individual is estimated to be 420 cubic centimetres.[20] We do not know the adult's brain size because a large portion of the skull is missing. *A. sediba* walked upright,[21] but it also had long arms and is thought to have been equally at home in the trees.[22] (Also see Appendix B.)

Paranthropus

The australopithecines and *Homo* lived at the same time as a hominin from a different genus. *Paranthropus* comprises a group of species dated to between one and 2.5 million years ago. They, too, lived only in Africa.

The species that entered our story in chapter 1 is *Paranthropus boisei*. It lived between 2.3 and 1.3 million years ago in the regions we now call Tanzania, Kenya and Ethiopia.[23]

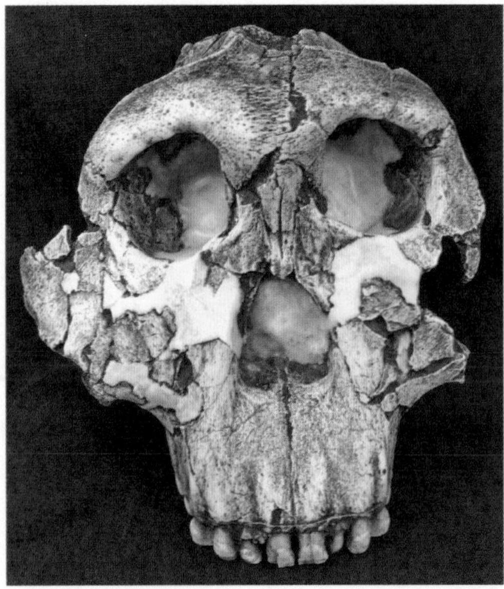

P. boisei, replica model at the School of Archaeology and Anthropology, ANU. Image provided by Debbie Argue.

The *P. boisei* females are estimated to have stood 124 centimetres tall, with the males slightly taller at 137 centimetres. Their faces projected outwards and were very wide and dished, and there was a large crest on the skull that anchored the massive chewing muscles. It had huge cheek teeth (molars) and a large jaw, earning it the nickname 'Nutcracker Man'. Its brain size varied between 475 and 545 cubic centimetres.[24] It had great upper-limb strength and probably spent time climbing trees. Its hand structure would have enabled it to make stone tools, but it could not grip with its thumb and forefinger together as we can (called the 'precision grip').[25]

P. boisei was called *Australopithecus* (*Zinjanthropus*) *boisei* for some time, a name that appears in older publications. Hence, it is also nicknamed 'Zinj'. (Also see Appendix B.)

Homo

The species of *Homo* differ from the australopithecines in their larger brain sizes, less projecting faces, and smaller jaws relative to their skull size. An early member of this species, *H. habilis*, was similar in stature to the australopithecines and also had long arms compared to their legs. But from about 1.5 million years ago, arm-to-leg ratios in *Homo* were the same as ours, and brain sizes increased.

Originally, tool-making was considered a defining characteristic of *Homo*, a feature that differentiated it from the australopithecines. However, we now know that australopithecines made tools too.[26]

The earliest evidence of *Homo* is a partial jaw, with most of the teeth intact, unearthed in January 2013 in the Ledi-Geraru research area in Afar Regional State in Ethiopia. It has the museum name of LD 350-1, but people usually just call it 'Ledi-Geraru'. It is dated to between 2.80 and 2.75 million years ago, and even though it is officially in *Homo*, it is considered a transitional form that existed between *Australopithecus* and *Homo*.[27] This is because it combines the primitive characteristics seen in the australopithecines with traits observed in early *Homo*.

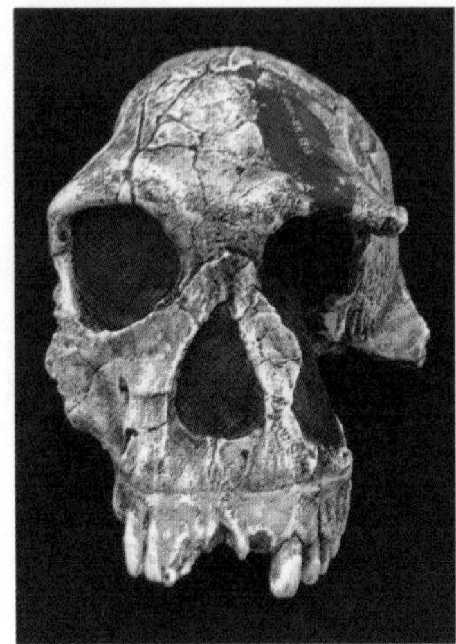

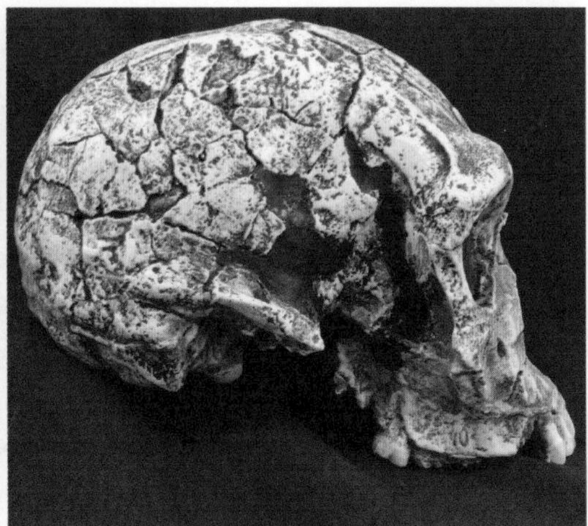

Left: *H. habilis*, KNM-ER 1813, replica model at the School of Archaeology and Anthropology, ANU, front view. Image provided by Debbie Argue.
Right: *H. habilis*, KNM-ER 1813, replica model at the School of Archaeology and Anthropology, ANU, side view. Image provided by Debbie Argue.

Homo habilis

Homo habilis is known from the Koobi Fora research area of Kenya and Olduvai Gorge in Tanzania. The species is dated from between 2.35 and 1.65 million years ago.[28]

H. habilis individuals would have stood 100–135 centimetres tall. Their brain sizes exceeded those of the australopithecines, ranging from 500 to 687 cubic centimetres.[29] *H. habilis* skulls were somewhat more rounded than the australopithecines', their faces projected less, and the ridges above the eyes were not as pronounced. Their jaws were also smaller.[30] (Also see Appendix B.)

Between 1.4 and 1.8 million years ago, *H. habilis* and *Paranthropus* shared the Koobi Fora region with another species, *H. ergaster*, while in Eurasia, in Georgia, another hominin flourished: *H. georgicus*. Soon afterwards, 1–1.8 million years ago (the dates are somewhat unclear), we have *H. erectus* way over in Java.

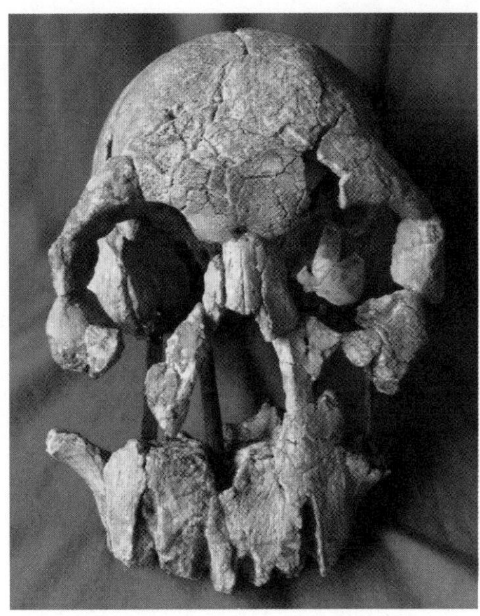

H. rudolfensis, KNM-ER 1470, original, curated at the National Museums, Nairobi, Kenya. Image provided by Debbie Argue and Colin Groves.

Homo rudolfensis

Homo rudolfensis is a species of early *Homo* that is known from approximately 1.8 million years ago in East Africa. We have one skull and two upper leg bones of this species,[31] but despite more than forty years of research and analyses, it is still unclear where this hominin sits on the human evolutionary tree. Some researchers do not consider *H. rudolfensis* to be a separate species and would include the fossils in *H. habilis*.

Homo erectus

Homo erectus was discovered in 1891 on the island of Java. The first find, a molar tooth, was excavated 64 kilometres from the village of Trinil, on the banks of the Solo River. A month later, a skullcap was excavated near the location of the molar. Soon after, an upper leg bone was discovered. There is now an assortment of *H. erectus* skulls and teeth, along with four (perhaps five[32]) upper-leg bones, from the Trinil and Sangiran areas of Java. (Also see Appendix B.)

Recently, half a world away from Java, a two-million-year-old partial skull of a juvenile was discovered during excavations in Drimolan cave

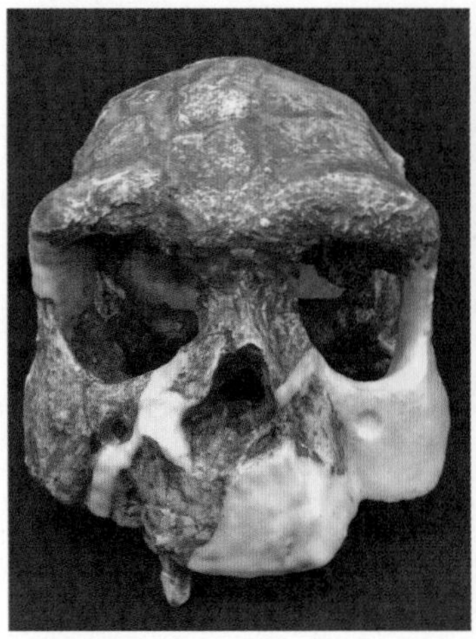
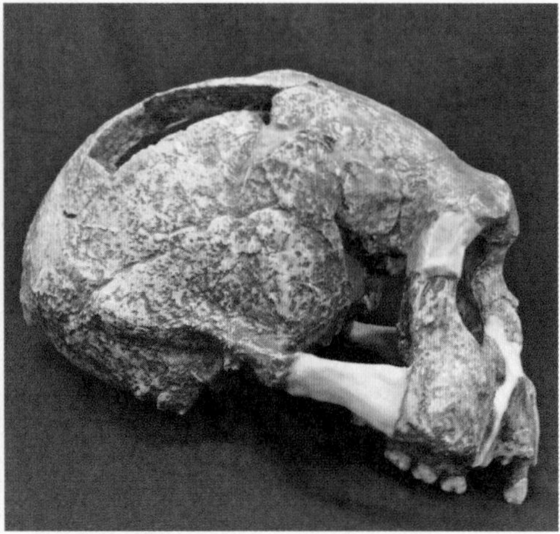

Left: *H. erectus*, Sangiran 17, replica model at the School of Archaeology and Anthropology, ANU, front view. Image provided by Debbie Argue.
Right: *H. erectus*, Sangiran 17, replica model at the School of Archaeology and Anthropology, ANU, side view. Image provided by Debbie Argue.

in South Africa. It is very similar to a juvenile *H. erectus* skullcap from Mojokerto in Java and has been attributed to *H. erectus*.[33]

A plethora of attempts have been made to nail down the dates of the Java fossils, a task made difficult by the geology of the Sangiran region. Nonetheless, this has yielded a range of proposed dates: around 1.8 million years ago,[34] between 1.5 and 1.02 million years ago,[35] between 1.3 and 1.5 million years ago,[36] and 800 000 years ago.[37]

H. erectus is also known from China, although some researchers attribute those fossils to a different species, *H. pekinensis*. This is because they differ somewhat in skull shape and some other characteristics. They have been dated to either between 460 000 and 230 000 years ago, or to between 550 000 and 300 000 years ago,[38] making them much younger than *H. erectus* from Trinil and Sangiran.

H. erectus' brain size ranges from 813 to 1059 cubic centimetres,[39] considerably larger than *H. habilis* brains. The skulls vary from massively thick to more gracile (thin) forms and are relatively long and low. They have what we might call 'super-structures': elongated, raised mounds of bone in various places. There is one quite low down across the back of the skull (occipital torus) and one that lies along the top of the skull (sagittal torus). Stretching across the face and projecting like a visor above the eyes is a supraorbital torus. Behind this there is a narrow, valley-like form before the forehead rises back gently (unlike our vertical forehead). The jaw is fairly massive and has no chin, instead sloping backwards from below the lower front teeth.

The only Sangiran *H. erectus* with substantial facial features preserved is called Sangiran 17. Its face is very wide, with the cheekbones flaring out beyond the rest of the face. In side view, the face projects.

Homo ergaster

The fossils that would later become *Homo ergaster* were discovered in the East Rudolf area of Kenya in 1971 by Richard Leakey. Four years later, Colin Groves and Vratislav Mazák were studying the whole suite of fossils that had been discovered in South Africa: the australopithecines, *H. habilis* and others. Using fossil measurement data from various publications, they patiently performed a myriad calculations. And out of this massive piece of work, something interesting emerged. A set of *H. habilis* remains from the 1.4–1.8-million-year-old level in the East Rudolf area did not fit with *H. habilis* at all. The fossil teeth were smaller and wider than *H. habilis'* teeth, and they were even smaller than the teeth of *A. africanus*; the tooth roots were different too. Groves and Mazák named these fossils *Homo ergaster*.[40]

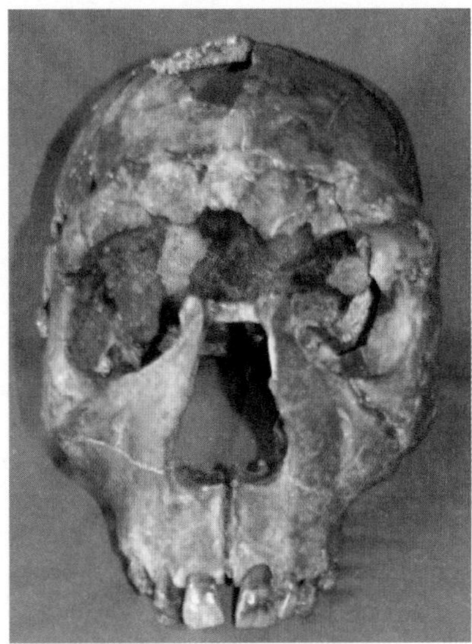

H. ergaster, KNM-WT 15000, original, curated at the National Museums, Nairobi, Kenya. Image provided by Debbie Argue and Colin Groves.

Almost a decade later, in 1984, an impressive skeleton was discovered and nicknamed 'Nariokotome Boy'. It is generally included in this species, although opinions differ. If this individual is *H. ergaster*, we can say that the species walked upright and it had a modern arm–leg ratio, rather than the long arms of *H. habilis*. The form of Nariokotome Boy's shoulder, however, was primitive. Unlike ours, its shoulders were hunched, and the arm sockets faced forward such that it could not rotate its arms as we can.[41] This configuration also occurs in *H. georgicus*, *H. floresiensis* and *H. naledi*.

On how they chose the name for this species, Groves and Mazák explained: 'Under normal circumstances it would be customary to honour the discoverer (i.e. R. E. F. Leakey) by naming a new form, which he had declined to name; but the name *leakeyi* is preoccupied in the genus *Homo*.'[42] By 'preoccupied' the researchers meant that *leakeyi* was already being used in a species name: *Homo leakeyi* is the name for a fossil skull from Olduvai Gorge (OH 9), named in 1963. Under the International Code of Zoological Nomenclature, species names can only be used once, which stands to reason. The name that Groves and Mazák settled on, '*ergaster*' is from the Greek word meaning 'workman'. It was chosen because stone tools were found in the same locality.[43]

Researchers do not unanimously accept this species, however, and the *H. ergaster* fossils are often incorporated into *H. erectus*. Some researchers prefer the term 'African *Homo erectus*' to make it clear they are talking about the fossils from Africa and not *H. erectus* from Java or China. But 'African *Homo erectus*' is a confusing term. Not only is it not a species name, frustratingly, researchers differ in the fossils they include in the group.

Homo georgicus

While the lives of *H. habilis* and *Paranthropus* were in full swing in East Africa, further north in what is now the nation of Georgia, another species appeared. Fossils discovered between 1991 and 1999 at the archaeological site of Dmanisi were formally declared a new species, *Homo georgicus*, in 2002.[44] The remains now comprise five skulls; four jaws, with various teeth intact; backbones, ribs and shoulder bones; an upper and a lower limb; and a partial foot.

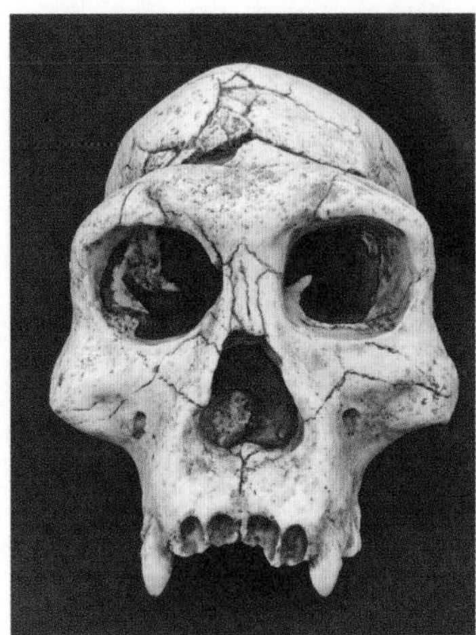

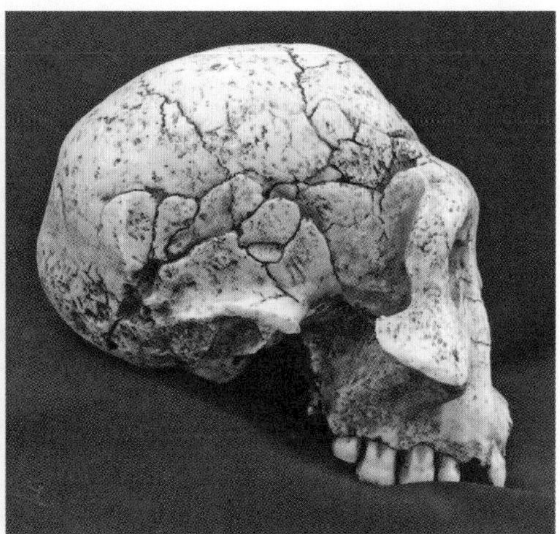

Left: *H. georgicus*, D2700, replica model at the School of Archaeology and Anthropology, ANU, front view. Image provided by Debbie Argue.
Right: *H. georgicus*, D2700, replica model at the School of Archaeology and Anthropology, ANU, side view. Image provided by Debbie Argue.

The fossils from Dmanisi are puzzling because they vary so widely in form compared to most other hominin species. Yet, except for one of the jaws, the fossils were excavated from a thin layer of a large, deep excavation. That layer, and the fossils within it, have been dated to 1.85 million years ago. The aforementioned jaw is from a different layer and is possibly older than the other fossils.[45]

H. georgicus walked upright and had the same arm–leg ratio as modern humans. But in other respects they were more primitive. For example, the shoulder blade shows that their shoulders were hunched and faced forward, as we have seen for *H. ergaster*.

It is actually rare to see the Dmanisi fossils referred to as *H. georgicus*. Under the International Code of Zoological Nomenclature, the first name awarded to a species is the one that has to be used for it,[46] yet frequently we see the Dmanisi fossils called *H. erectus*. *H. georgicus* was of course named in honour of the country in which the fossils were discovered. (Also see Appendix B.)

The mysterious Denisovans

Almost unbelievably, we came to know about the Denisovans courtesy of the tip of a finger bone no bigger than a coffee bean.[47] Professor Michael Shunkov, co-director with Professor Anatoly Derevianko (Institute of Archaeology and Ethnography in Novosibirsk, the Siberian branch of the Russian Academy of Sciences) of excavations in Denisova Cave in the Altai region of Siberia,[48] discovered the bone there in 2008.[49] There was nothing about the finger bone that seemed unusual. The archaeologists simply didn't know if it came from a modern human or a Neanderthal, so it was sent to the Max Planck Institute for Evolutionary Anthropology, a leading player in the extraction of Neanderthal DNA. The result was astonishing. This unassuming little bone came from an individual belonging to a completely unknown group of hominins,[50] who were soon called Denisovans. The bone is dated to 30 000–48 000 years ago.[51]

The identification of a new species from DNA was an enormous scientific breakthrough. This finger bone and its DNA must surely go down as one of the most outstanding discoveries in human evolution of the twenty-first century.

And the DNA had more to reveal. It showed that the Denisovans were a sister group to the Neanderthals; that is, the two groups shared a common ancestor, from whom they diverged around 640 000 years ago.[52] After the two groups split, they had separate population histories: the Neanderthals lived in Europe and Western Asia, while the Denisovans probably inhabited Asia. The mysterious Denisovans got around, though. They contributed their genes to the ancestors of present-day populations of island South-East Asia and Oceania, being genetically closer to people from New Guinea than to anyone on mainland Eurasia. To a lesser extent, their genes have been found in people in the Americas and across mainland Asia.[53]

Although the Denisovans and Neanderthals largely lived in separate regions, they clearly overlapped in the Altai area, as evidenced by the discovery of bones from both groups in Denisova Cave.

In one case, at least, the overlap became personal. Viviane Slon and colleagues found that a fragment of long bone (arm or leg) from a female, labelled Denisova 11, had equal amounts of DNA from a Denisovan and a Neanderthal.[54] Interbreeding between two species? It seemed unlikely. To ensure they were right about this, Slon and colleagues repeated the analysis six times, each time using fresh samples of DNA from the bone. They obtained the same results each time: Mum was a Neanderthal and Dad was a Denisovan.[55] And there was more. Slon and her colleagues also found that the Denisovan father had some Neanderthal admixture in *his* DNA.[56] Were the Denisovans and Neanderthals interbreeding as a matter of course? Perhaps, but there is not yet enough evidence to make a definitive call on this.

In 1980, 2500 kilometres away from Denisova Cave, a Buddhist monk saw an unusual-looking partial jawbone with two intact teeth when he entered Baishiya Karst Cave, on the Tibetan Plateau in Xiahe County, Gansu Province, China, to pray.[57] The jaw was eventually curated at Lanzhou University, also in Gansu. Dr Frido Welker (University of Copenhagen) recaps what happened next:

> August 2016, and Dongju Zhang of Lanzhou University (China), myself, and Jean-Jacques Hublin (Max Planck Institute for Evolutionary Anthropology) had a meeting in our offices in Leipzig. Dongju showed us tantalizing photos of an unpublished mandible from the Tibetan Plateau, China. Not quite Neanderthal, not really *Homo erectus*, and definitely not belonging to our species, *Homo sapiens*. Where did it fit in?[58]

The jaw was dated to 160 000 years ago. The researchers tried analysing ancient DNA to identify the species. They found that no DNA had been preserved, but what was present were proteins, which contain genetic information. The team compared these with the corresponding proteins from fossils of Neanderthals, Denisovans, modern humans and apes. There was a match with a Denisovan.[59] It appears that this jaw represents the first substantial piece of the Denisovan skeleton.

To live at such a high altitude, the Denisovans had to have successfully adapted to relatively low oxygen levels.[60] Modern-day high-altitude Tibetans have physiologically adapted to this environment through a particular gene.

And, as it happens, it was the Denisovans who contributed this gene to the ancestors of the Tibetans.[61]

Tantalisingly, Denisovans also have DNA from an unknown archaic hominin. Could this be another invisible hominin group for which we as yet have no fossil evidence? Or could it be from a species we know about but for which we have no DNA? Perhaps a hominin fossil with Denisovan DNA is lying unobtrusively in a museum collection somewhere.

Homo naledi

In 2013, a whole cache of bones was discovered 30 metres underground in the Dinaledi and Lesedi chambers of a cave in the Rising Star system outside Johannesburg in South Africa.[62] The 1500 bones excavated to date are from all parts of the body: the head, jaw, teeth, backbone, shoulder, ribs, pelvis, an arm, leg, hand and foot, representing fifteen adults and youngsters. They have been dated to 241 000–335 000 years ago.[63] Even though *H. sapiens* were around for part of this time (albeit a long way away, in Morocco), the Rising Star remains are not *H. sapiens*. The bones have a puzzling array of primitive and modern characteristics, especially for a species that lived so recently.[64] (Also see Appendix B.)

Homo naledi's brain was australopithecine-sized, at 460–560 cubic centimetres.[65] Its body had an australopithecine-like barrel-shaped trunk, and a similar pelvis structure. It also had the hunched, archaic shoulder form in which the arms face more forward than ours, as we have seen for *H. ergaster*, *H georgicus* and *H. floresiensis*.[66] *H. naledi*'s fingers were long and even more curved than those of the australopithecines, showing that they could climb and suspend from trees,[67] yet these hands could also manipulate objects, as we can.[68] Standing at 144.5–147.8 centimetres, *H. naledi* was as tall as small-bodied modern human groups. Its lower leg, foot and ankle were human-like as well, and like us it could stride along on two feet.

This species is an enigma in the way in which its bones somehow accumulated in two deep, unconnected chambers 30 metres below the surface, where no light shines. These were not one-off accidental or catastrophic events. Rather, 236 000–335 000 years ago, intact bodies were deliberately disposed of in these two chambers. Was *H. naledi* bravely going down the

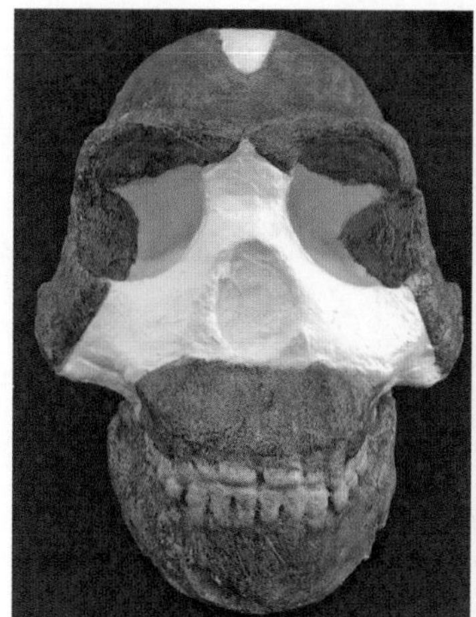

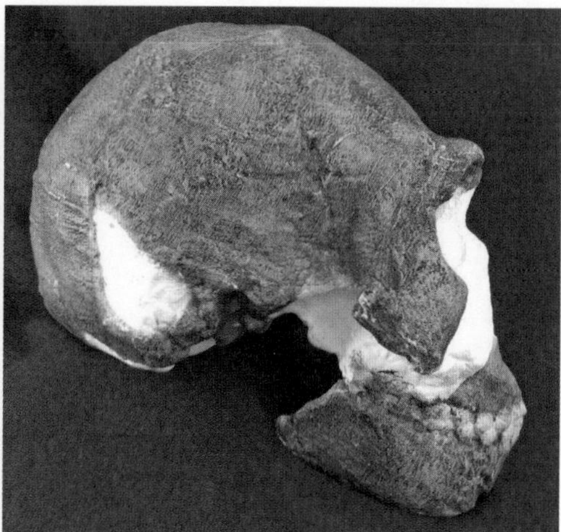

Left: *H. naledi*, replica model, School of Archaeology and Anthropology, ANU, front view. The areas of white represent filling for missing bone. Image provided by Debbie Argue.
Right: *H. naledi*, replica model, School of Archaeology and Anthropology, ANU, side view. The areas of white represent filling for missing bone. Image provided by Debbie Argue.

steep, narrow, dark passages carrying bodies with them to deposit? Were *H. sapiens* also in the locality, and equally bravely taking *H. naledi* bodies into the two dark chambers? Or were there other openings, as yet not identified, into which these individuals were thrown?[69]

Homo sapiens

There is only one kind of human on earth today: us. We live on most landmasses and in almost all environments. And we are getting older by the minute. When I was first studying human evolution, we were 190 000 years old. Then we were 230 000 years old. Recently, however, remains found at the archaeological site of Jebel Irhoud in Morocco were dated to 315 000 years ago. These early *Homo sapiens* fossils include two almost-complete skulls, two braincases, a nearly complete jaw and the fossils of some juveniles.[70]

Appendix B

Hominin fossil discoveries

Accounts of the discoveries of hominin fossils show just how diverse these experiences can be.

Homo erectus

From his childhood in the mid-nineteenth century, the anatomist Dr Eugene Dubois had been fascinated with the idea of an evolutionary 'missing link' between apes and us (the concept of a 'missing link' is now outdated, of course, because we know that human evolution is more like a branching bush than a straight line from an ape-like ancestor to modern humans). Dubois was busily working in Holland as an anatomist, lecturer and physician when the discovery of two strange human-like fossil skeletons in a cave in Spy, Belgium was announced in 1887. The remains were subsequently identified as *Homo neanderthalensis*.[1] Not only were the skeletons different from ours, they were found with the fossilised bones of extinct rhinos, mammoth, reindeer, horses and cave bears.[2] These were ice-age animals; the two Neanderthal fossils, however, were of indeterminate age.

Dubois was both fascinated and galvanised. Human evolutionary discoveries were happening in his time! He re-read Ernst Haeckel's *History of Creation*, in which this scholar predicted that one day an ape-man fossil would be found, a hypothetical 'missing link' between the apes and man which he called *Pithecanthropus*. Dubois' ambition was set: he would find the 'missing link'.

There was no point in looking in Europe as everything was too young—there were no known apes there. In any case, everyone else was searching that region. Charles Darwin had predicted that Africa was the go for early evidence of human evolution, but Dubois thought that gorillas and chimps were too different from us and their ancestors wouldn't fit as a missing link. Asia, though, had gibbons. It had orangutans. The known animal fossils were of Pleistocene age. Asia, decided Dubois, was where he should go. However, his applications for scientific funding were gently rejected. So he joined the army, obtaining a posting as a physician, and headed for the Dutch East Indies (now Indonesia). Once there, when time allowed, he explored for possible sites. He found some animal fossils which, although they were scrappy, impressed his superior officer enough that he assigned two army engineers, Sergeant G Kreile and Sergeant A De Winter,[3] as well as some local labour, to help. Dubois secured this support by pronouncing that if the Dutch didn't look for important fossils in this region, other nations would.[4]

According to a story told in 1885 by a local villager, the bones of giants had been found near the village of Trinil. The inhabitants reported this to the garrison commander at Ngawi, who informed Dubois' superiors. In turn, he was charged with carrying out archaeological excavations.[5] The work commenced almost immediately, and in August 1891 the men found fossils of extinct stegodonts, buffalo and deer.[6] The following month, they excavated the upper molar of an ape-like primate from the left bank of the Solo River. This was a very promising find, but within another month came the prize: a fossilised skullcap, about the size of a large coconut. The skullcap was from something that had been big-brained, but it had a chimpanzee-like brow ridge, behind which was a shelf-like form, after which the forehead sloped back. Dubois compared this skullcap with all known fossil apes: no match. It was not an ape, nor was it a Neanderthal, and it was certainly not a modern human.[7] The following year, Dubois' men found a leg bone 15 metres upstream from where the skullcap had been found, and in the same archaeological level. The form of the leg bone showed that this hominin had walked upright.

At first Dubois attributed the skullcap and molar to a new species of chimpanzee: *Anthropopithecus erectus Dubois*.[8] The species name *erectus* referred to it being an upright walker. Fairly soon, however, he recognised that this was no chimpanzee, and he formally used Haeckel's hypothetical appellation for the genus of the new species, naming it *Pithecanthropus erectus*.[9] We can see his change of mind in a manuscript he prepared (see next page).

Pithecos means ape and *anthropus* means man. Together, they mean 'ape-human that stands upright'. Today, this species is considered to be a member of *Homo*: *Homo erectus*.

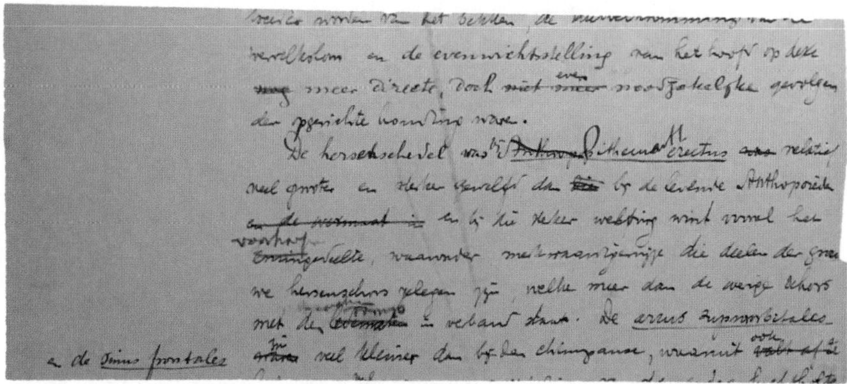

Dr Eugene Dubois manuscript. Image kindly provided by Jon de Vos.

Australopithecus africanus

It was probably not the incentive of a £5 prize but the sheer excitement of discovery that led Miss Josephine Salmons to eagerly tell her anatomy lecturer, Professor Raymond Dart, that she'd come across something interesting.[10] The year was 1924, and Dart was in his second year of teaching in the Medical School at the University of Witwatersrand in Johannesburg, South Africa. He needed to set up an anatomy museum so that the students could learn about the bones of any and every animal, so he'd offered a prize to the student who collected the most interesting finds during their holidays.

Miss Salmon had noticed a fossil skull on the mantel at a family friend's place. It was a baboon fossil from the Northern Lime Company's mine at Taung, and a rarity, thought Dart. But even better, Miss Salmons casually remarked to Dart that fossil skulls and bones were often turning up at the mine. Through a friend of a friend, Dart arranged for boxes of stone blocks containing bone fragments to be sent to him.[11]

The boxes turned up just as Dart was donning formal attire as the best man at a wedding. The inopportune timing caused consternation in his knowing wife, who asked Dart to please not go delving into 'all that rubble' until after the wedding was over.[12] But Dart couldn't resist those boxes. As soon as his wife left the room, he wrenched open the lid of the first box:

nothing of interest. In the second box, though, on top of the rock heap, was a brain cast from the inside of an ape-like skull—one that was too big to be from an ape. 'Was there anywhere, among this pile of rocks, a face to fit the brain?' thought Dart. At this moment the bridegroom came in and beseeched him: 'My God Ray you've got to finish dressing—or I'll have to find another best man.'[13]

As soon as the last guests had left, Dart, whose fledging Medical School lab had few appropriate tools, ducked out to a hardware shop to buy a hammer and chisel to remove the rock encasing the brain cast. Once he'd exposed enough of the skull to get an idea of its location in the rock, he borrowed his wife's knitting needles to gently prise the matrix away.[14] The job took weeks—it took until the seventy-third day for the rock to part. What appeared was a child's face, one with its first permanent teeth emerging.[15] This was no ape but some sort of unknown hominin. 'I doubt if there was any parent prouder of his offspring that I was of my Taungs baby that Christmas of 1924,' wrote Dart.[16]

And so *Australopithecus africanus* became known to the world, the first fossil hominin to be discovered in Africa. But its tiny brain did not fit with conventional views of human evolution, according to which an increase in brain size was paramount. It took nearly twenty years for the scientific community to accept it.

Since then, many more fossil skulls, jaws, teeth and body bones of *A. africanus* have been discovered in South Africa.

Homo habilis

Louis and Mary Leakey had been working at Olduvai Gorge, Tanzania for a few years when, in 1955, they were joined by a Mrs Brown, followed by a Miss Trace and a Miss Goodall.[17] Two years later, among a large number of fossil animal bones, this team found two hominid teeth. One was a child's deciduous molar, but it was very large compared to that of the australopithecines, and it had a complex pattern of lumps (cusps) on its top surface. The other tooth was a deciduous canine (eyetooth).[18] Over the next few years, other mystery fossils turned up: some foot bones, a few hand bones, some tiny skull fragments, a couple of shoulder bones and a partial hominid mandible, the latter found by the Leakeys' eldest son, Jonathan. All these fossils were parts of individuals who lived at the same time as the australopithecines, but they did not match anything in that genus. The Leakeys realised that they were dealing with a quite distinctive early hominid.[19]

In 1963, even more fossils were discovered that were not australopithecines, including a thick hominid skull,[20] which was discovered by Miss M Cropper. The new finds fitted in with the previous mystery finds.[21] Putting it all together in 1964, Louis Leakey and the palaeoanthropologist Philip Tobias and the primatologist John Napier identified these fossils as comprising a new species, *Homo habilis*.[22] The species name *habilis* was taken from the Latin word meaning 'able, handy, mentally skilful, vigorous'. This name was suggested by Professor Raymond Dart and reflects the idea that the species made and used the stone tools that were found in the same area.

Paranthropus

Mary Leakey discovered *Paranthropus boisei* in 1959 whilst working in Olduvai Gorge. She and Louis had turned their attention to the oldest level in the gorge, called Bed 1. Helson Mukiri had the first stroke of luck when he discovered a hominid tooth. Even better, when the limestone matrix was removed, the tooth was found to actually be in a jaw. Meanwhile, a documentary team scheduled to film one of the Leakeys' excavations was delayed. This gave the Leakeys a couple more days in which to keep looking in Olduvai Gorge. Louis, however, fell ill with flu, so Mary headed out by herself. As she recollects,

> so it came about that on the morning of 17 July I went out by myself with the two dalmatians Sally and Victoria, to see what I could find of interest at the nearby Bed 1 exposures ... One bone that caught and held my eye was not lying loose on the surface but projecting from beneath ... It had a hominid look but the bones seemed enormously thick—too thick, surely. I carefully brushed away a little of the deposit, and then I could

> see parts of two large teeth in place in an upper jaw. They *were* hominid. It was a hominid skull, apparently still *in situ*, and there was a lot of it there. I rushed back to camp to tell Louis, who leaped out of bed, and then we were soon back at the site, looking at my find together. Louis was sad that the skull was not of early *Homo*, but he concealed his feelings well and expressed only mild disappointment. *Zinjanthropus* had come into our lives.[23]

Louis created the new genus *Zinjanthropus* because he saw that the skull was not a straightforward member of *Australopithecus*. *Zinj* is derived from an ancient Arabic word for East Africa. The Leakeys chose the species name *boisei* in gratitude to Charles Boise, their very generous supporter.

Mary and Louis invited Philip Tobias, who had succeeded Raymond Dart as Chair of Anatomy at the University of Witwatersrand, to undertake a formal study of the skull. Tobias concluded that it did actually belong in the australopithecines and named it *Australopithecus boisei*.[24] However, it is now considered to be part of the genus *Paranthropus* and thus is called *Paranthropus boisei*.

Australopithecus afarensis

The year was 1974 and the temperature was approaching 43°C in the Hadar region of Ethiopia. The researchers working there hadn't found much in the way of fossils under that gruelling sun.

'I've had it,' said graduate student Tom Gray. 'When do we head back to camp?'

'Right now,' replied Don Johanson, co-leader of the fieldwork team. 'But let's go back this way and search the bottom of that little gully over there.'

There was virtually no bone in the gully, but as the pair turned to leave, Johanson noticed something lying on the ground partway up the slope: a hominin arm bone!

'Can't be. It's too small. It has to be a monkey bone,' said Gray.

'Hominid,' was Johanson's response.

'What makes you so sure?'

'That piece right next to your hand.'

Tom Gray picked it up. It was the back of a small skull. A few feet away lay a leg bone, a couple of vertebrae, part of a pelvis and some ribs.

Could they be part of a primitive skeleton? Johanson knew nothing like this had been found anywhere else.[25]

That afternoon, everyone was in that gully, searching, searching, searching. Johanson recalls the excitement on that first night:

> The camp was rocking with excitement ... we never went to bed at all. We talked and talked. We drank beer after beer. There was a tape recorder in the camp, and a Beatles song 'Lucy in the Sky with Diamonds' went belting out into the night sky, and was played at full volume over and over again out of sheer exuberance. At some point ... I don't remember when, the new fossil picked up the name of Lucy, and has been known so ever since, although its proper acquisition number is AL 288-1.[26]

This species is now represented by the remains of more than 300 individuals.

Homo georgicus

The modern-day village of Dmanisi in the Caucasus country of Georgia lies within a picturesque rural landscape near the meeting of two rivers. Although rather quiet and small now, during the Middle Ages Dmanisi was one of the most important cities along the old Silk Road. Excavating near the village, archaeologists found that an ancient citadel contained large, deep grain-storage pits. One of these pits took on new significance when, in 1983, palaeontologist Absalom Vekua of the Georgian Academy of Sciences discovered the tooth of a long-extinct rhinoceros in it.[27] It appeared that Dmanisi was going to reveal much about times far earlier than the medieval period.

Sure enough, with the reinvigorated archaeological focus, stone tools were revealed in the excavations, suggesting that early hominins had lived in this area. Then, in 1991, the first hominin fossil was discovered. It was a jawbone, found lying underneath the skeleton of a sabre-toothed cat. The researchers estimated the jaw to be around 1.8–1.6 million years old.[28] No hominins were thought to have ventured out of Africa this early, so the discovery was therefore greeted with scepticism when it was announced at an international meeting of experts: it was the wrong place for an ancient fossil hominin, and in any case, the jaw was in too pristine a condition to be particularly old.[29]

Validation of the Georgian researchers' claim would come eight years later. Archaeology student Gotcha Kiladze was showing a group of schoolchildren around the Dmanisi archaeological site, just before the season's excavations were to commence, when he spotted the glint of a bone poking out from the sediments. It turned out to be the skull of a hominin. Two months later, another student unearthed a second skull.[30] Since then, three more skulls and several body parts have been discovered at the site.

Australopithecus sediba

It all began with a nine-year-old boy discovering another nine-year-old boy. It was just that the second nine-year-old was around 1.9 million years old.

Palaeoanthropology Professor Lee Berger of Witwatersrand University had been looking for fossil bones in the so-called Cradle of Humankind in South Africa's Gauteng province for many, many years, but he hadn't found much of anything. So he turned to Google Earth, which enabled him to identify an area of about 500 caves in a region called Malapa,[31] 50 kilometres from Johannesburg. Scientists had as yet explored none of these caves.

On 15 August 2013, Berger, his colleague Job Kibii and Berger's nine-year-old son Matthew headed out to explore Malapa. When they came across a cave in which they saw bones, Berger said, 'Let's look around.' Matthew ran off and within a few minutes he called out to his dad. He'd found a fossil in a largish rock, a shoulder bone—a clavicle. Turning the rock over, Berger and Kibii could see a jawbone and a tooth. 'We knew we were onto something—we'd hit the jackpot,' recalls Job. Sure enough, more of the 'jackpot' lay waiting for them in that cave: a skull along with part of a body protruding from a rock.

In fact, they'd found two bodies close together: a boy estimated to be nine years old and an adult female. It was a treasure trove: between the two individuals, nearly all the bones of the body were represented. These fossils are nearly two million years old.[32]

Homo naledi

Thirty miles north-west of Johannesburg is a cave called Rising Star that has been a magnet for cavers since the 1960s. Its network of channels and caverns has been well mapped. But in 2013, two cavers, Steven Tucker and Rick Hunter, nonetheless hoped to find as-yet-unexplored areas of the cave system. After squeezing through some very narrow chutes, they eventually dropped into a cavity and saw bones everywhere.[33] They lost no time in informing Professor Lee Berger who, taking one look at the photos the cavers showed him, realised that all of his other projects would have to take a back seat.

Such a narrow access route to the chamber meant that a select group of archaeologists would need to be recruited. Dr Elen Feuerriegel, at the time a PhD candidate in the School of Archaeology and Anthropology at the ANU, recalls:

> I first saw the ad on a microblogging website called Tumblr ... Detail was sparse about the nature of the job but the required qualifications were very specific: applicants had to have a Masters or PhD in palaeontology, archaeology or a similar field, field experience plus caving or climbing experience, the ability to work as a team in claustrophobic spaces, and we had to be of reasonably small stature and build.
>
> The critical piece was this: we had to be able to fit through an 18-cm-wide slot. Bizarre! And intriguing—certainly intriguing enough that I decided to throw my hat into the ring. I had all the required qualifications: a Masters in biological anthropology and I had just started

a PhD in palaeoanthropology. I had caving experience. I had excavated fossil hominin material in caves in Spain. Plus I am 5'3' and skinny.

I didn't have any real expectation that my application would be considered. If I had all the right qualifications and experience, surely many other people would be more qualified! So you can imagine my surprise when a few days after I had sent in my CV, I get an email from Lee Berger himself: would I be willing to sit for a Skype interview tomorrow? Baffled and excited, I wrote back confirming the time and my contact details for the interview. I spent the rest of the day preparing, including setting up an obstacle course in my apartment with an 18-cm pinch, just to make sure I could do the job.

Within a few hours of finishing my interview with Lee, I got another email: be ready to fly to South Africa within a week. I didn't know it, but it was the start of the most exciting and rewarding five years of my life.[34]

Elen was one of six females recruited from this worldwide quest. She and the other archaeologists—Alia Gurtov (University of Wisconsin), Marina Elliott (MA from Simon Fraser University, Canada), Becca Peixotto (American University, Washington, DC), K Lindsay Eaves Hunter (University of Witwatersrand) and Hannah Morris (University of Georgia)—became known as the 'Underground Astronauts' because they were connected remotely with Lee Berger and the rest of the team via a camera and phone system.

A total of 1550 bones from all parts of the body, and representing at least fifteen individuals, have been uncovered in the Dinaledi Chamber of the cave, while excavations within the Lesedi Chamber have revealed a further 131 bones representing three individuals.[35] The bones from both chambers are from immature and mature individuals. The mystery question: how did these bones accumulate deep within this cave complex?

Glossary

Archaic character or characteristic

A feature on a bone that is found in an ancestral species.

Character or characteristic

Any attribute of an organism; in palaeoanthropology it refers to a physical feature on a bone.

Character state

A specified variation in a character.

Cladistics

The principle of deducing the interrelationships of organisms based on character states.

Corpus

In human biology, the main part or body of bone.

Hominin and hominid

'Hominin' refers to a group consisting of modern humans, extinct species and all our immediate ancestors, including the australopithecines, *Paranthropus* and *Ardipithecus*. We use it to designate the line leading to modern humans and the various members of the human evolutionary tree. 'Hominid' refers to a group comprising all modern and extinct great apes—modern humans, chimpanzees, gorillas orangutans, and their immediate ancestors. The word used to have the same meaning as hominin has now, so be aware that when older textbooks or journal articles use 'hominid', they are referring to 'hominin'. The name switch came about when changes were made to the way humans, gorillas and orangutans were classified.[1]

Hypothesis

An idea concerning an event and its possible explanation. It is usually based on evidence and its merit lies in the capacity to use it to make testable predictions.[2]

In situ

Situated in its original place. That is, it has not been moved from its original place of deposition by, for example, erosional action, or as a result of other events.

International Code of Zoological Nomenclature

A set of rules for naming and the use of names in zoology. These rules are governed by the International Commission on Zoological Nomenclature.

Lineage

A series of populations connected by a continuous line of descent from ancestor to descendant.

Miocene Epoch

The Miocene Epoch is dated from 23.03 million to 5.3 million years ago.

Model

Models in science are produced to show possible relationships among two or more factors, and to shed light on what might have happened as a result of these relationships.[3]

Palaeoanthropology

'Palaeo' means old, especially in relation to the geological past, and 'anthropology' is the study of humans. So 'palaeoanthropology' is the study of the evolutionary origins and development of the human family. It often involves the scientific study of human fossils.

Palaeoanthropologist

A person who specialises in the study of human evolutionary origins.

Pleistocene Epoch

The epoch dated from 2.58 million to 12 000 years ago. Early Pleistocene: 2.58 million to 780 000 years ago; Middle Pleistocene: 780 000 to 12 600 years ago; Late Pleistocene: 12 600 to 12 000 years ago.[4]

Primitive character or characteristic

A feature on a bone that is found in an ancestral species.

Strata

Series of layers of sediment or other naturally occurring material deposited over time.

Taxon

A taxon is a group of one or more populations of an organism that is considered by taxonomists to form a unit. They are arranged in a hierarchical order from kingdoms to subspecies. The plural is taxa. Genus and species are taxonomic categories.

Taxonomy

The theory and practice of classifying organisms; the branch of biology that classifies all living things.

Notes

Prologue

1 Years later, a friend told me that it was this book that led her to archaeology, too.
2 I passed.
3 Helgen, Kristofer M, Eulogy for Colin Peter Groves, personal communication, 7 December 2017.
4 Rookmaaker, K, and J Robovský, 'Bibliography of Colin Peter Groves (1942–2017), an Anthropologist and Mammalian Taxonomist', *Lynx (Prague)*, no. 49, 2018, pp. 255–94.
5 Helgen, KM, personal communication.
6 I thank Myron Shekelle for providing me with this information.
7 Unpublished nomination of Colin Groves for the International Primatological Society Lifetime Achievement Award, 2016.
8 Cave, Christine, personal communication, 9 June 2021.

1 The discovery

1 Brown, P, et al., 'A New Small-Bodied Hominin from the Late Pleistocene of Flores, Indonesia', *Nature*, vol. 431, no. 7012, 2004, pp. 1055–61.
2 Morwood, M, and P van Oosterzee, *The Discovery of the Hobbit: The Scientific Breakthrough that Changed the Face of Human History*, Random House, Milsons Point, NSW, 2007, p. 7.
3 Morwood and van Oosterzee, p. 54.
4 Ibid., p. 41.
5 Sutikna, T, interview, 7 December 2018.
6 Morwood and van Oosterzee, pp. 68–9.

7 Technically, the dates are between 61 ± 13 thousand and 45 ± 9 thousand years ago: Roberts, RG, R Jones and MA Smith, 'Thermoluminescence Dating of a 50 000-Year-Old Human Occupation Site in Northern Australia', *Nature*, vol. 345, no. 6271, 1990, pp. 153–6.
8 Roberts, B, personal communication, 16 July 2019; Swisher, CC, et al., 'Latest *Homo erectus* of Java: Potential Contemporaneity with *Homo sapiens* in Southeast Asia', *Science*, vol. 274, no. 5294, 1996, pp. 1870–4.
9 Roberts, B, personal communication.
10 Ibid.
11 Now Associate Professor at Macquarie University.
12 Morwood and van Oosterzee, p. 92.
13 Sutikna, T, interview.
14 Jatmiko, personal communication, 31 January 2019.
15 Ibid.
16 Orluc, L, *Man from Flores: The Tale of the Last Hobbits*, Ex Nihilo/France Televisions, 52 minutes, 2011.
17 Brown et al., 'A New Small-Bodied Hominin from the Late Pleistocene of Flores, Indonesia'.
18 Jungers, WL, et al., 'Descriptions of the Lower Limb Skeleton of *Homo floresiensis*', *Journal of Human Evolution*, vol. 57, no. 5, 2009, pp. 538–54; Larson, SG, et al., 'Descriptions of the Upper Limb Skeleton of *Homo floresiensis*', *Journal of Human Evolution*, vol. 57, no. 5, 2009, pp. 555–70.
19 Jungers et al., 'Descriptions of the Lower Limb Skeleton of *Homo floresiensis*'.
20 Now Professor Peter Brown.
21 More information about the species mentioned in this book may be found in Appendix A.
22 Brown et al., 'A New Small-Bodied Hominin from the Late Pleistocene of Flores, Indonesia'.
23 Kubo, D, RT Kono and Y Kaifu, 'Brain Size of *Homo floresiensis* and Its Evolutionary Implications', *Proceedings of the Royal Society B-Biological Sciences*, vol. 280, no. 1760, 2013, p. 1.
24 Tocheri, MW, et al., 'The Primitive Wrist of *Homo floresiensis* and Its Implications for Hominin Evolution', *Science*, vol. 317, no. 5845, 2007, pp. 1743–5. Also see Orr, Caley M, et al., 'New Wrist Bones of *Homo floresiensis* from Liang Bua (Flores, Indonesia)', *Journal of Human Evolution*, vol. 64, 2013, pp. 109–29.
25 *The Hobbit Enigma*, written, directed and produced by Simon Nasht and Annamaria Talas, Essential Media and Entertainment, and ScreenSound Australia, developed and produced in association with the Australian Broadcasting Corporation, 2008.
26 Lieberman, DE, '*Homo floresiensis* from Head to Toe', *Nature*, vol. 459, no. 7243, 2009, pp. 41–2.
27 Jungers, W, and K Baab, 'The Geometry of Hobbits: *Homo floresiensis* and Human Evolution', *Significance*, vol. 6, 2009, pp. 159–64.

28 Larson, SG, et al., '*Homo floresiensis* and the Evolution of the Hominin Shoulder', *Journal of Human Evolution*, vol. 53, no. 6, 2007, pp. 718–31.

29 Lordkipanidze, D, et al., 'Postcranial Evidence from Early Homo from Dmanisi, Georgia', *Nature*, vol. 449, no. 7160, 2007, pp. 305–10.

30 Feuerriegel, EM, et al., 'The Upper Limb of *Homo naledi*', *Journal of Human Evolution*, vol. 104, 2017, pp. 155–73.

31 Larson, S, personal communication, 2019.

32 Falk, D, et al., 'The Brain of LB1, *Homo floresiensis*', *Science*, vol. 308, no. 5719, 2005, pp. 242–5.

33 Sutikna, T, et al., 'Revised Stratigraphy and Chronology for *Homo floresiensis* at Liang Bua in Indonesia', *Nature*, vol. 532, no. 7599, 2016, pp. 366–9.

34 Sutikna, T, et al., 'The Spatio-Temporal Distribution of Archaeological and Faunal Finds at Liang Bua (Flores, Indonesia) in Light of the Revised Chronology for *Homo floresiensis*', *Journal of Human Evolution*, vol. 124, 2018, pp. 52–74.

35 Van den Bergh, GD, et al., 'The Liang Bua Faunal Remains: A 95 K.yr. Sequence from Flores, East Indonesia', *Journal of Human Evolution*, vol. 57, no. 5, 2009, pp. 527–37.

36 Van den Bergh, GD, et al., 'The Youngest Stegodon Remains in Southeast Asia from the Late Pleistocene Archaeological Site Liang Bua, Flores, Indonesia', *Quaternary International*, vol. 182, 2008, pp. 16–48, information on pp. 34–5.

37 Moore, MW, et al., 'Continuities in Stone Flaking Technology at Liang Bua, Flores, Indonesia', *Journal of Human Evolution*, vol. 57, no. 5, 2009, pp. 503–26.

38 Sutikna et al., 'The Spatio-Temporal Distribution of Archaeological and Faunal Finds at Liang Bua (Flores, Indonesia) in Light of the Revised Chronology for *Homo floresiensis*'.

39 Leakey, M, *Disclosing the Past*, Weidenfeld and Nicholson, London, 1984, pp. 120, 127.

40 Napier, J, 'Fossil Hand Bones from Olduvai Gorge', *Nature*, vol. 196, no. 4853, 1962, pp. 409–11.

41 Leakey, LSB, JR Napier and PV Tobias, 'New Species of Genus *Homo* from Olduvai Gorge', *Nature*, vol. 202, no. 492s, 1964, pp. 5–7. The Latin word *habilis* means 'able, handy, mentally skilful, vigorous'. For more information about this species, see Appendix A.

42 Smith, GK, 'Large Cave Discoveries on Flores (Indonesia), July–August 2006', Australian Speleological Federation Biennial Conference, Mount Gambier, South Australia, 6–12 January 2007.

43 Smith, GK, 'Large Cave Discoveries on Flores (Indonesia)', *Caves Australia, Journal of The Australian Speleological Federation Inc.*, vol. 172, 2006, pp. 1–19.

44 Hantoro, W, personal communication, 2019. Also see Hantoro, W, 'Occupation and Diaspora of Austronesia: Learning from Geo-Oceanoclimatology Perspective in Indonesian Maritime Island on Enhancing Resilience Living in the Coastal Plain and Small Island', International Symposium on Austronesian

Diaspora, in *Austronesian Diaspora: A New Perspective*, Gajah Mada University Press, Yogyakarta, 2016.

45 Gagan, M, interview, 21 November 2018.

46 All artefacts and bones from Liang Bawah are archived at ARKENAS.

47 Gagan, MK, et al., 'Geoarchaeological Finds below Liang Bua (Flores, Indonesia): A Split-Level Cave System for *Homo floresiensis*?', *Palaeogeography Palaeoclimatology Palaeoecology*, vol. 440, 2015, pp. 533–50.

48 Smith, GK, personal communication, 2019.

49 Ibid.

50 Sutikna, T, interview, 7 December 2018.

51 St Pierre, EJ, et al., 'Preliminary U-Series and Thermo-Luminescence Dating of Excavated Deposits in Liang Bua Sub-Chamber, Flores, Indonesia', *Journal of Archaeological Science*, vol. 40, no. 1, 2013, pp. 148–55.

52 My colleague Associate Professor Julien Louys says that they used to be called whip spiders and are now called whip scorpions or amblypygi. They are tailless.

2 Controversy from the start

1 *The Hobbit Enigma*, written, directed and produced by Simon Nasht and Annamaria Talas, Essential Media and Entertainment, and Screensound Australia, developed and produced in association with the Australian Broadcasting Corporation, 2008.

2 Brown, P, et al., 'A New Small-Bodied Hominin from the Late Pleistocene of Flores, Indonesia', *Nature*, vol. 431, no. 7012, 2004, pp. 1055–61.

3 No *H. erectus* fossils have been discovered on Flores.

4 Henneberg, M, and J Schofield, *The Hobbit Trap*, Wakefield Press, Adelaide, 2008, pp. 35–6.

5 Ibid., p. 29.

6 Ibid., p. 38.

7 Poulianos, AN, 'An Early Minoan Microcephale', *Anthropos*, vol. 2, 1975, pp. 42–7.

8 Darwin, Charles, *The Descent of Man, and Selection in Relation to Sex*, Gibson Square Books, London, 2003 [1871].

9 Stringer, C, *The Origin of Our Species*, Penguin Books, London, 2012, pp. 8–12.

10 Henneberg, M, and A Thorne, 'Flores Human May Be a Pathological *Homo sapiens*', *Before Farming*, vol. 4, 2004, pp. 1–4.

11 See Jacob, T, *Some Problems Pertaining to the Racial History of the Indonesian Region: A Study of Human Skeletal and Dental Remains from Several Prehistoric Sites in Indonesia and Malaysia*, Drukkerij Neerlandia, Utrecht, 1967.

12 Woods, CG, J Bond and W Enard, 'Autosomal Recessive Primary Microcephaly (MCPH): A Review of Clinical, Molecular, and Evolutionary Findings', *American Journal of Human Genetics*, vol. 76, no. 5, 2005, pp. 717–28.

13 Böök, JA, JW Schut and SC Reed, 'A Clinical and Genetic Study of Microcephaly', *American Journal of Mental Deficiency*, vol. 57, 1953, pp. 637–43.

14 Mochida, GH, and CA Walsh, 'Molecular Genetics of Human Microcephaly', *Current Opinion in Neurology*, vol. 14, no. 2, 2001, pp. 151–6.

15 Ibid.
16 Poulianos, 'An Early Minoan Microcephale'.
17 Ibid.
18 Ibid.
19 OH 24 is a *H. habilis* skull from Olduvai Gorge, Tanzania ('OH' refers to 'Olduvai Hominid'). The other skull was KNM-ER 3733 ('KNM' refers to Kenya National Museum; 'ER' refers to East Rudolf).
20 We assumed the leg bone was from *A. garhi*, but we have since been alerted to the fact that there is some question about this.
21 Argue, D, et al., '*Homo Floresiensis*: Microcephalic, Pygmoid, *Australopithecus*, or *Homo*?', *Journal of Human Evolution*, vol. 51, no. 4, 2006, pp. 360–74. The news of our work raced around the world in a flurry of publicity.
22 Falk, D, FE Lepore and A Noe, 'The Cerebral Cortex of Albert Einstein: A Description and Preliminary Analysis of Unpublished Photographs', *Brain*, vol. 136, 2013, pp. 1304–27.
23 Falk, D, et al., 'Brain Shape in Human Microcephalics and *Homo floresiensis*', *PNAS*, vol. 104, no. 7, 2007, pp. 2513–18.
24 Martin, RD, et al., 'Flores Hominid: New Species or Microcephalic Dwarf?', *Anatomical Record Part A: Discoveries in Molecular Cellular and Evolutionary Biology*, vol. 288a, no. 11, 2006, pp. 1123–45.
25 Ibid., p. 1141.
26 Richards, GD, 'Genetic, Physiologic and Ecogeographic Factors Contributing to Variation in *Homo sapiens*: *Homo floresiensis* Reconsidered', *Journal of Evolutionary Biology*, vol. 19, no. 6, 2006, pp. 1744–67.
27 Hershkovitz, I, L Kornreich and Z Laron, 'Comparative Skeletal Features between *Homo floresiensis* and Patients with Primary Growth Hormone Insensitivity (Laron Syndrome)', *American Journal of Physical Anthropology*, vol. 136, no. 3, 2008, p. 373.
28 Laron, Z, 'Lessons from 50 Years of Studying Laron Syndrome', *Endocrine Practice*, vol. 21, no. 12, 2015, pp. 1395–402.
29 Ibid.
30 Laron, Z, 'Laron Syndrome (Primary Growth Hormone Resistance or Insensitivity): The Personal Experience 1958–2003', *The Journal of Clinical Endocrinology & Metabolism*, vol. 89, no. 3, 2004, pp. 1031–44.
31 Ibid.
32 Laron, 'Lessons from 50 Years of Studying Laron Syndrome'.
33 Falk, D, et al., 'The Type Specimen (LB1) of *Homo floresiensis* Did Not Have Laron Syndrome', *American Journal of Physical Anthropology*, vol. 140, no. 1, 2009, pp. 52–63.
34 See also Hershkovitz, Kornreich and Laron, 'Comparative Skeletal Features between *Homo floresiensis* and Patients with Primary Growth Hormone Insensitivity (Laron Syndrome)'.
35 Obendorf, PJ, CE Oxnard and BJ Kefford, 'Are the Small Human-Like Fossils Found on Flores Human Endemic Cretins?', *Proceedings of the Royal Society*

B: Biological Sciences, vol. 275, no. 1640, 2008, pp. 1287–96; Oxnard, CE, PJ Obendorf and BJ Kefford, 'Post-Cranial Skeletons of Hypothyroid Cretins Show a Similar Anatomical Mosaic as *Homo floresiensis*', *Plos One*, vol. 5, no. 9, 2010, e13018.

36 Colin Groves and Catherine Fitzgerald's conference presentation was called 'Healthy Little Hobbits or Victims of Sauron?'

37 Brown, P, 'LB1 and LB6 *Homo floresiensis* Are Not Modern Human (*Homo sapiens*) Cretins', *Journal of Human Evolution*, vol. 62, no. 2, 2012, pp. 201–24.

38 Baab, KL, KP McNulty and K Harvati, '*Homo floresiensis* Contextualized: A Geometric Morphometric Comparative Analysis of Fossil and Pathological Human Samples', *Plos One*, vol. 8, no. 7, 2013, e69119.

39 This photo was taken as part of the project 'The Flores hobbit: *Homo floresiensis* or microcephalic eastern Indonesian?', which was financed by an Australian Research Council Discovery Grant.

40 Vannucci, RC, TF Barron and RL Holloway, 'Craniometric Ratios of Microcephaly and LB1, *Homo floresiensis*, Using MRI and Endocasts', *Proceedings of the National Academy of Sciences of the United States of America*, vol. 108, no. 34, 2011, pp. 14043–8.

41 Henneberg, M, et al., 'Evolved Developmental Homeostasis Disturbed in LB1 from Flores, Indonesia, Denotes Down Syndrome and Not Diagnostic Traits of the Invalid Species *Homo floresiensis*', *Proceedings of the National Academy of Sciences of the United States of America*, vol. 111, no. 33, 2014, pp. 11967–72.

42 Baab, KL, et al., 'A Critical Evaluation of the Down Syndrome Diagnosis for LB1, Type Specimen of *Homo floresiensis*', *Plos One*, vol. 11, no. 6, 2016, e0155731.

43 Westaway, MC, et al., 'Mandibular Evidence Supports *Homo floresiensis* as a Distinct Species', *Proceedings of the National Academy of Sciences of the United States of America*, vol. 112, no. 7, 2015, pp. E604–E605.

44 Schwartz, JH, and I Tattersall, 'The Human Chin Revisited: What Is It and Who Has It?', *Journal of Human Evolution*, vol. 38, no. 3, 2000, pp. 367–409.

45 Balzeau, A, and P Charlier, 'What Do Cranial Bones of LB1 Tell Us about *Homo floresiensis*?', *Journal of Human Evolution*, vol. 93, 2016, pp. 12–24.

46 More recently, dates of between 73 000 and 63 000 years ago have been proposed from archaeological work in Sumatra: see Westaway, KE, et al., 'An Early Modern Human Presence in Sumatra 73 000–63 000 years ago', *Nature*, vol. 548, no. 7667, 2017, pp. 322–5.

47 Lee, MS, personal communication, 2020.

48 Tucci, S, et al., 'Evolutionary History and Adaptation of a Human Pygmy Population of Flores Island, Indonesia', *Science*, vol. 361, no. 6401, 2018, pp. 511–15.

3 Fitting *Homo floresiensis* on the family tree

1 Although we blithely use the term 'recent common ancestor' in cladistics, it is somewhat frustrating that cladistics cannot tell us if (say) fossil X is an actual ancestor of any species Y. We cannot use cladistics to find out which species was the ancestor of *H. floresiensis.* Encouragingly, our colleague Professor Mike Lee, a cladistics expert, says that new methods are currently being developed to identify species' ancestors: Lee, MS, personal communication, 2020.

2 Our genus is *Homo*—more on this in Appendix A.

3 Out of Africa 2 is when *Homo sapiens* moved into Eurasia.

4 Brown, P, et al., 'A New Small-Bodied Hominin from the Late Pleistocene of Flores, Indonesia', *Nature*, vol. 431, no. 7012, 2004, pp. 1055–61.

5 Morwood, M, et al., 'Further Evidence for Small-Bodied Hominins from the Late Pleistocene of Flores, Indonesia', *Nature*, vol. 437, 13 October 2005, doi:10.1038/nature04022; Morwood, MJ, and WL Jungers, 'Conclusions: Implications of the Liang Bua Excavations for Hominin Evolution and Biogeography', *Journal of Human Evolution*, vol. 57, no. 5, 2009, pp. 640–8.

6 Argue, D, et al., '*Homo floresiensis*: A Cladistic Analysis', *Journal of Human Evolution*, vol. 57, no. 5, 2009, pp. 623–39.

7 *A. afarensis* and *H. floresiensis* have the same pelvis form; *H. floresiensis*' upper-leg length is akin to that of *A. afarensis*; and the front of the *H. floresiensis* jaws, and its body proportions, are similar to *A. afarensis*.

8 Argue, D, et al., 'The Affinities of *Homo floresiensis* Based on Phylogenetic Analyses of Cranial, Dental, and Postcranial Characters', *Journal of Human Evolution*, vol. 107, 2017, pp. 107–33.

9 Brown et al., 'A New Small-Bodied Hominin from the Late Pleistocene of Flores, Indonesia'.

10 Dembo, M, et al., 'Bayesian Analysis of a Morphological Supermatrix Sheds Light on Controversial Fossil Hominin Relationships', *Proceedings of the Royal Society B: Biological Sciences*, vol. 282, no. 1812, 2015, pp. 133–41.

11 The *H. erectus* calvaria Zeitoun and colleagues used were Trinil, Sangiran 2 and Sangiran 17. *H. ergaster* was represented by KNM-WT 15000. Zeitoun, V, V Barriel and H Widianto, 'Phylogenetic Analysis of the Calvaria of *Homo floresiensis*', *Comptes Rendus Palevol*, vol. 15, no. 5, 2016, pp. 555–68, Table 1 and Figure 3.

12 Technical note: Consistency Index 0.386; Retention Index 0.426; Bootstrap value not provided because it is less than 50 per cent. The test that potentially supports the result is Bremer support, which is 4. Ibid., p. 560.

13 Ibid., p. 564.

14 Ibid.

15 Argue, D, et al., '*Homo floresiensis*: Microcephalic, Pygmoid, *Australopithecus*, or *Homo*?', *Journal of Human Evolution*, vol. 51, no. 4, 2006, pp. 360–74.

16 Baab, KL, KP McNulty and K Harvati, '*Homo floresiensis* Contextualized: A Geometric Morphometric Comparative Analysis of Fossil and Pathological Human Samples', *Plos One*, vol. 8, no. 7, 2013, e69119.

17 Skull D2700.

18 Gordon, AD, L Nevell and B Wood, 'The *Homo floresiensis* Cranium (LB1): Size, Scaling, and Early Homo Affinities', *Proceedings of the National Academy of Sciences of the United States of America*, vol. 105, no. 12, 2008, pp. 4650–5.
19 Lyras and colleagues referred to these fossils as *H. erectus*.
20 Lyras, GA, et al., 'The Origin of *Homo floresiensis* and Its Relation to Evolutionary Processes under Isolation', *Anthropological Science*, vol. 117, no. 1, 2009, pp. 33–43.
21 Brown et al., 'A New Small-Bodied Hominin from the Late Pleistocene of Flores, Indonesia'.
22 Kaifu, Y, et al., 'Craniofacial Morphology of *Homo floresiensis*: Description, Taxonomic Affinities, and Evolutionary Implication', *Journal of Human Evolution*, vol. 61, no. 6, 2011, pp. 644–82.
23 Ibid.
24 Kaifu, Y, et al., 'Unique Dental Morphology of *Homo floresiensis* and Its Evolutionary Implications', *Plos One*, vol. 10, no. 11, 2015, e0141614.
25 Brown et al., 'A New Small-Bodied Hominin from the Late Pleistocene of Flores, Indonesia'.
26 LH 4 is the museum catalogue number for Laetoli Hominid 4 jaw.
27 Brown, P, and T Maeda, 'Liang Bua *Homo floresiensis* Mandibles and Mandibular Teeth: A Contribution to the Comparative Morphology of a New Hominin Species', *Journal of Human Evolution*, vol. 57, no. 5, 2009, pp. 571–96.
28 Ibid., p. 571.
29 Argue et al., '*Homo floresiensis*: Microcephalic, Pygmoid, *Australopithecus*, or *Homo*?'
30 Brown and Maeda, 'Liang Bua *Homo floresiensis* Mandibles and Mandibular Teeth: A Contribution to the Comparative Morphology of a New Hominin Species': because *H. floresiensis* is more primitive than *H. georgicus*.
31 *Dental Perspectives on Human Evolution: State of the Art in Dental Paleoanthropology*, in *Vertebrate Paleobiology and Paleoanthropology*, Max Planck Institute subseries in human evolution, Springer, Dordrecht, 2007.
32 And retained in living apes.
33 Brown and Maeda, 'Liang Bua *Homo floresiensis* Mandibles and Mandibular Teeth: A Contribution to the Comparative Morphology of a New Hominin Species', p. 580.
34 Also referred to as *H. pekinensis*: see Appendix A.
35 Falk, D, et al., 'The Brain of LB1, *Homo floresiensis*', *Science*, vol. 308, no. 5719, 2005, pp. 242–5.
36 Ibid., p. 245.
37 Jungers, W, and K Baab, 'The Geometry of Hobbits: *Homo floresiensis* and Human Evolution', *Significance*, vol. 6, 2009, pp. 159–64.
38 Brown and Maeda, 'Liang Bua *Homo floresiensis* Mandibles and Mandibular Teeth: A Contribution to the Comparative Morphology of a New Hominin Species'; Morwood and Jungers, 'Conclusions: Implications of the Liang Bua Excavations for Hominin Evolution and Biogeography'.

39 The navicular bone.
40 Bennett, MR, et al., 'Early Hominin Foot Morphology Based on 1.5-Million-Year-Old Footprints from Ileret, Kenya', *Science*, vol. 323, no. 5918, 2009, pp. 1197–201.
41 Jungers and Baab, 'The Geometry of Hobbits: *Homo Floresiensis* and Human Evolution'.
42 Jungers, WL, et al., 'The Foot of *Homo floresiensis*', *Nature*, vol. 459, no. 7243, 2009, pp. 81–4. A 'pygmy', in anthropology, is a member of any human group whose adult males grow to less than 150 centimetres in average height: see *Encyclopaedia Britannica*, 'Pygmy', 11 October 2019, https://www.britannica.com/topic/Pygmy (viewed November 2021).
43 Lee, MS, personal communication, 2020.
44 Diniz-Filho, JAF, and P Raia, 'Island Rule, Quantitative Genetics and Brain-Body Size Evolution in *Homo floresiensis*', *Proceedings of the Royal Society B: Biological Sciences*, vol. 284, no. 1857, 2017.
45 Ibid., p. 6.
46 Scardia, G, et al., 'Chronologic Constraints on Hominin Dispersal outside Africa since 2.48 Ma from the Zarqa Valley, Jordan', *Quaternary Science Reviews*, vol. 219, 2019, pp. 1–19.
47 Zhu, ZY, et al., 'Hominin Occupation of the Chinese Loess Plateau since about 2.1 Million Years Ago', *Nature*, vol. 559, no. 7715, 2018, pp. 608–12; Dennell, R, and W Roebroeks, 'An Asian Perspective on Early Human Dispersal from Africa', *Nature*, vol. 438, no. 7071, 2005, pp. 1099–104.
48 Herries, AIR, et al., 'Contemporaneity of *Australopithecus*, *Paranthropus*, and early *Homo erectus* in South Africa', *Science*, vol. 368, no. 47, 2020.
49 Scardia, G, et al., 'What Kind of Hominin First Left Africa?', *Evolutionary Anthropology*, vol. 30, 2021, pp. 122–7.
50 Skull D2280 is 650 cubic centimetres, skull D2700 is 600 cubic centimetres, and skull D2822 is 780 cubic centimetres. The brain sizes of *H. habilis* range from 509 to 687 cubic centimetres.
51 The brain sizes of *H. ergaster* range from 750 to 848 cubic centimetres.
52 Hall, R, 'Sundaland and Wallacea: Geology, Plate Tectonics and Palaeogeography', in KGJ David, J Gower, James E Richardson, Brian R Rosen, Lukas Rüber and Suzanne T Williams (eds), *Biotic Evolution and Environmental Change in Southeast Asia*, Cambridge University Press, Cambridge, 2012.

What's in a name?

1 The history of the peer review concerning the naming of the species is by an editor of *Nature*, Henry Gee, in his review of Morwood and Van Oosterzee's 2007 book *The Discovery of the Hobbit*: see *Nature*, vol. 446, 26 April 2007, pp. 979–80.

4 Fossil bones of the So'a Basin, Flores

1 Fijn, Natasha, *Exploring Lost Caves in Flores Indonesia*, Fijn Film Productions, Canberra, 2011, https://vimeo.com/manage/videos/89161775 (viewed November 2021).

2 Van den Bergh, GD, 'The Late Neogene Elephantoid-Bearing Faunas of Indonesia and Their Palaeozoogeographic Implications', *Scripta Geologica*, vol. 117, 1999, Natuur en Boek, Leiden, p. 266.

3 Aziz, F, MJ Morwood and GD van den Bergh, *Pleistocene Geology, Palaeontology and Archaeology of the So'a Basin, Central Flores, Indonesia*, Pusat Survei Geologi, Bandung Geologi, 2008, p. 4.

4 Van den Bergh, 'The Late Neogene Elephantoid-Bearing Faunas of Indonesia and Their Palaeozoogeographic Implications', p. 266.

5 Aziz, Morwood and van den Bergh, *Pleistocene Geology, Palaeontology and Archaeology of the Soa Basin, Central Flores, Indonesia*, p. 3.

6 Van den Bergh, GD, personal communication, 31 October 2021, from information in Knepper, GM, *Floresmens, het leven van Theo Verhoeven missionaris en archaeoloog* (*Floresman, the Life of Theo Verhoeven, Missionary and Archaeologist*), Boekscout, the Netherlands, 2019.

7 De Vos, J, personal communication, 22 September 2019.

8 Hooijer, DA, 'A Stegodon from Flores', *Treubia*, vol. 24, 1958, pp. 119–33.

9 Maringer and Verhoeven, 'Recent Discovery of a Palaeolithic Past in Flores, Indonesia and Its Contribution to the Research of Most Ancient Southeast Asia', p. 248.

10 Ibid., pp. 247–63.

11 Van den Bergh, 'The Late Neogene Elephantoid-Bearing Faunas of Indonesia and Their Palaeozoogeographic Implications', pp. 286–328.

12 Maringer and Verhoeven, 'Recent Discovery of a Palaeolithic Past in Flores, Indonesia and Its Contribution to the Research of Most Ancient Southeast Asia', p. 248.

13 Knepper, *Floresmens, het leven van Theo Verhoeven missionaris en archaeoloog* (*Floresman, the Life of Theo Verhoeven, Missionary and Archaeologist*), p. 270.

14 Maringer and Verhoeven, 'Recent Discovery of a Palaeolithic Past in Flores, Indonesia and Its Contribution to the Research of Most Ancient Southeast Asia', pp. 261–2.

15 Morwood, MJ, et al., 'Stone Artefacts from the 1994 Excavation at Mata Menge, West Central Flores, Indonesia', *Australian Archaeology*, vol. 44, no. 1, 1997, pp. 26–34.

16 Van den Bergh, G, personal communication, 2020.

17 Ibid.

18 Meijer, HJM, personal communication.

19 *Pak* is the appropriate way to address adult males of a particular age or status in most parts of Indonesia. A Bahasa Indonesia word, *pak* is essentially equivalent to the English noun 'mister'. As with the English equivalent, *pak* can be used on

its own as a form of address or used as a title (Mr) before a person's name. In the latter case, however, *pak* usually precedes an individual's first name, rather than their last; for example, in Indonesia it would be most common to say 'Mr John' rather than 'Mr Smith' or 'Mr John Smith', as would be the case in English. I thank an anonymous reviewer for this information.

20 Van den Bergh, G, personal communication, 9 August 2020.

21 From the Geological Survey Institute, Bandung, Java.

22 The Netherlands Foundation for Tropical Research and the Geological Research and Development Centre at Bandung, Indonesia funded this research: see Sondaar, PY, et al., 'Middle Pleistocene Faunal Turnover and Colonization of Flores (Indonesia) by *Homo-erectus*', *Comptes Rendus De L Academie Des Sciences Serie Ii*, vol. 319, no. 10, 1994, pp. 1255–62.

23 Hnatyshin, D, and VA Kravchinsky, 'Paleomagnetic Dating: Methods, MATLAB Software, Example', *Tectonophysics*, vol. 630, 2014, pp. 103–12.

24 Ghaleb, B, et al., 'Timing of the Brunhes-Matuyama Transition Constrained by U-Series Disequilibrium', *Scientific Reports*, vol. 9, 2019.

25 Van den Bergh, G, personal communication, 2020.

26 Sondaar et al., 'Middle Pleistocene Faunal Turnover and Colonization of Flores (Indonesia) by *Homo-erectus*'.

27 Aziz, Morwood and van den Bergh, *Pleistocene Geology, Palaeontology and Archaeology of the Soa Basin, Central Flores, Indonesia*, p. 11.

28 Sondaar et al., 'Middle Pleistocene Faunal Turnover and Colonization of Flores (Indonesia) by *Homo-erectus*'; van den Bergh, GD, et al., 'Did *Homo erectus* Reach the Island of Flores?', *Indo-Pacific Prehistory Association Bulletin*, vol. 14 (Chiang Mai Papers), no. 1, 1996, pp. 27–36.

29 Aziz, Morwood and van den Bergh, *Pleistocene Geology, Palaeontology and Archaeology of the Soa Basin, Central Flores, Indonesia*, p. 12.

30 Morwood, MJ, et al., 'Fission-Track Ages of Stone Tools and Fossils on the East Indonesian Island of Flores', *Nature*, vol. 392, no. 6672, 1998, pp. 173–6.

31 Aziz, Morwood and van den Bergh, *Pleistocene Geology, Palaeontology and Archaeology of the Soa Basin, Central Flores, Indonesia*, pp. 3–14.

32 Brumm, A, personal communication, 2020. Flakes are small stone tools that have been struck off a larger stone.

33 Ibid. Fluvial strata are layers of sediment deposited by water action over the years.

34 Brumm, A, et al., 'Hominins on Flores, Indonesia, By One Million Years Ago', *Nature*, vol. 464, no. 7289, 2010, pp. 748–52 and methods.

35 Van den Bergh, GD, et al., 'The Liang Bua Faunal Remains: A 95 K.yr. Sequence from Flores, East Indonesia', *Journal of Human Evolution*, vol. 57, no. 5, 2009, pp. 527–37.

36 Van den Bergh, G, personal communication, 31 October 2021.

37 Van den Bergh, GD, et al., 'Taphonomy of *Stegodon loresis* Remains from the Early Middle Pleistocene Archaeological Site Mata Menge, Flores, Indonesia', in DS Kostopoulos, E Vlachos and E Tsoukala (eds), *Scientific Annals of the School of Geology*, Aristotle University of Thessalonika, Thessalonika, 2014.

38 Van den Bergh, G, personal communication, 2020.
39 Calloway, E, 'Hobbit Relatives Hint at Family Tree', *Nature*, vol. 534, 2016, pp. 164–5.
40 Van den Bergh, GD, et al., '*Homo floresiensis*–Like Fossils from the Early Middle Pleistocene of Flores', *Nature*, vol. 534, no. 7606, 2016, pp. 245–59.
41 Brumm, A, et al., 'Age and Context of the Oldest Known Hominin Fossils from Flores', *Nature*, vol. 534, no. 7606, 2016, pp. 249–53.
42 Van den Bergh et al., '*Homo floresiensis*–Like Fossils from the Early Middle Pleistocene of Flores', see supplementary material.
43 Ibid.
44 Ibid.
45 Ibid., extended data Figure 5.
46 Ibid., extended data Figure 5 notes.
47 Ibid.
48 Ibid.
49 Ibid., Table 2.
50 Van den Bergh et al., '*Homo floresiensis*–Like Fossils from the Early Middle Pleistocene of Flores'; Kaifu, Y, et al., 'Unique Dental Morphology of *Homo floresiensis* and Its Evolutionary Implications', *Plos One*, vol. 10, no. 11, 2015, e0141614.
51 Van den Bergh et al., '*Homo floresiensis*–Like Fossils from the Early Middle Pleistocene of Flores', extended data Figure 2e.
52 Ibid., Figure 3.
53 Ibid., Figure 3b.

5 Float, walk or swim?

1 Van den Bergh, GD, 'The Late Neogene Elephantoid-Bearing Faunas of Indonesia and Their Palaeozoogeographic Implications', *Scripta Geologica*, vol. 117, 1999, Natuur en Boek, Leiden, p. 367.
2 Morwood, MJ, and WL Jungers, 'Conclusions: Implications of the Liang Bua Excavations for Hominin Evolution and Biogeography', *Journal of Human Evolution*, vol. 57, no. 5, 2009, pp. 640–8.
3 Ibid.
4 Dennell, RW, et al., 'The Origins and Persistence of *Homo floresiensis* on Flores: Biogeographical and Ecological Perspectives', *Quaternary Science Reviews*, vol. 96, 2014, pp. 98–107. Cyclones, however, form 7 degrees north and 25 degrees south of the equator and do not usually affect Indonesia: see Monk, KA, Y de Frietes and G Reksodiharjo-Lilley, *The Ecology of Nusa Tenggara and Maluku*, Periplus Editions, Hong Kong, 1997. There has been only one tropical cyclone in recent times, typhoon 'Vamei', which developed closer to the equator than any had previously done.
5 Dennell et al., 'The Origins and Persistence of *Homo floresiensis* on Flores: Biogeographical and Ecological Perspectives'.

6 Fan, WJ, et al., 'Variability of the Indonesian Throughflow in the Makassar Strait over the Last 30 Ka', *Scientific Reports*, vol. 8, 2018; Fallon, SJ, and TP Guilderson, 'Surface Water Processes in the Indonesian Throughflow as Documented by a High-Resolution Coral Delta Δ14C Record', *Journal of Geophysical Research-Oceans*, vol. 113, no. C9, 2008; Gordon, AL, RD Susanto and K Vranes, 'Cool Indonesian Throughflow as a Consequence of Restricted Surface Layer Flow', *Nature*, vol. 425, no. 6960, 2003, pp. 824–8; Mayer, B, and PE Damm, 'The Makassar Strait Throughflow and Its Jet', *Journal of Geophysical Research-Oceans*, vol. 117, 2012.

7 Sprintall, J, et al., 'Direct Estimates of the Indonesian Throughflow Entering the Indian Ocean: 2004–2006', *Journal of Geophysical Research-Oceans*, vol. 114, 2009.

8 Gordon, Susanto and Vranes, 'Cool Indonesian Throughflow as a Consequence of Restricted Surface Layer Flow'; Mayer and Damm, 'The Makassar Strait Throughflow and Its Jet'.

9 Gordon, Susanto and Vranes, 'Cool Indonesian Throughflow as a Consequence of Restricted Surface Layer Flow', Figure 1.

10 Brasseur, B, et al., 'Pedo-Sedimentary Dynamics of the Sangiran Dome Hominid Bearing Layers (Early to Middle Pleistocene, Central Java, Indonesia): A Palaeopedological Approach for Reconstructing "Pithecanthropus" (Javanese *Homo erectus*) Palaeoenvironment', *Quaternary International*, vol. 376, 2015, pp. 84–100.

11 Liu, T, and Z, Ding, 'Chinese Loess and the Paleomonsoon', *Annual Review of Earth and Planetary Sciences*, vol. 26, no. 1, 1998, pp. 111–45.

12 Falk, D, personal communication, 2 July 2019; Marshall, G, personal communication, 15 July 2019.

13 History.com, 'The Deadliest Tsunami in Recorded History', 2 October 2018, https://www.history.com/news/deadliest-tsunami-2004-indian-ocean (viewed November 2021).

14 International Tsunami Information Center, 'Summary of Earthquakes', July–September 2006, http://itic.ioc-unesco.org/images/docs/overview_17jul2006_Java.pdf (viewed November 2021).

15 Wikipedia, '2018 Sunda Strait Tsunami', 2 November 2021, https://en.wikipedia.org/wiki/2018_Sunda_Strait_tsunami#Casualties (viewed November 2021).

16 Hamzah, L, N Puspito and F Imamura, 'Tsunami Catalogue and Zones in Indonesia', *Journal of Natural Disaster Science*, vol. 22, no. 1, 2000, pp. 25–43.

17 Theil, M, and PA Haye, 'The Ecology of Rafting in the Marine Environment, III, Biogeographical and Evolutionary Consequences', in RN Gibson, RJA Atkinson and JDM Gordon (eds), *Oceanography and Marine Biology: An Annual Review*, vol. 44, 2006, pp. 323–431.

18 Heatwole, H, and R Levins, 'Biogeography of Puerto-Rican Bank: Flotsam Transport of Terrestrial Animals', *Ecology*, vol. 53, no. 1, 1972, pp. 112–17.

19 Theil and Haye, 'The Ecology of Rafting in the Marine Environment, III, Biogeographical and Evolutionary Consequences'.
20 Censky, EJ, K Hodge and J Dudley, 'Over-Water Dispersal of Lizards Due to Hurricanes', *Nature*, vol. 395, no. 6702, 1998, p. 556.
21 See Nuzula, F, ML Syamsudin and LPS Yuliadi, 'Eddies Spatial Variability at Makassar Strait–Flores Sea', *IOP Conference Series: Earth and Environmental Science*, 2017, Figure 2.
22 Ibid.
23 See Ruxton, GD, and DM Wilkinson, 'Population Trajectories for Accidental Versus Planned Colonisation of Islands', *Journal of Human Evolution*, vol. 63, no. 3, 2012, pp. 507–11; Dennell et al., 'The Origins and Persistence of *Homo floresiensis* on Flores: Biogeographical and Ecological Perspectives'.
24 BBC News, 'Tsunami "Miracle" Woman Pregnant', 6 January 2005.
25 Smith, JMB, 'Did Early Hominids Cross Sea Gaps on Natural Rafts?', in I Metcalf et al. (eds), *Faunal and Floral Migrations and Evolution in SE Asia–Australasia*, Swets and Zeitlinger, the Netherlands, 2001, pp. 409–16.
26 Piantadosi, CA, *The Biology of Human Survival: Life and Death in Extreme Environments*, Oxford University Press, Oxford, 2003.
27 Brown, P, et al., 'A New Small-Bodied Hominin from the Late Pleistocene of Flores, Indonesia', *Nature*, vol. 431, no. 7012, 2004, pp. 1055–61.
28 Drewnowski, A, CD Rehm and F Constant, 'Water and Beverage Consumption among Children Age 4–13y in the United States: Analyses of 2005–2010 NHANES Data', *Nutrition Journal*, vol. 12, 2013; Jequier, E, and F Constant, 'Water as an Essential Nutrient: The Physiological Basis of Hydration', *European Journal of Clinical Nutrition*, vol. 64, no. 2, 2010, pp. 115–23, Table 3.
29 Popkin, BM, KE D'Anci and IH Rosenberg, 'Water, Hydration, and Health', *Nutrition Reviews*, vol. 68, no. 8, 2010, pp. 439–58.
30 Jequier and Constant, 'Water as an Essential Nutrient: The Physiological Basis of Hydration', p. 120.
31 Hall, R, 'Sundaland and Wallacea: Geology, Plate Tectonics and Palaeogeography', in KGJ David, J Gower, James E Richardson, Brian R Rosen, Lukas Rüber and Suzanne T Williams (eds), *Biotic Evolution and Environmental Change in Southeast Asia*, Cambridge University Press, Cambridge, 2012.
32 Hall, R, 'The Palaeogeography of Sundaland and Wallacea since the Late Jurassic', *Journal of Limnology*, vol. 72, 2013, pp. 1–17, Figure 9.
33 Earle, W, 'On the Physical Structure and Arrangement of the Islands of the Indian Archipelago', *The Journal of the Royal Geographical Society of London*, vol. 15, 1845, pp. 358–65.
34 Wallace, AR, *The Malay Archipelago: The Land of the Orang-Utan, and the Bird of Paradise*, Oxford University Press, Oxford, 1986, pp. xiii–vi.
35 Van Wyhe, J, and K Rookmaaker, *Alfred Russel Wallace: Letters from the Malay Archipelago*, Oxford University Press, Oxford, 2013, pp. 296–301.

36 Wallace, AR, *My Life: A Record of Events and Opinions*, vol. 1, George Bell and Sons, London and Bombay, 1905, pp. 360, 403.
37 Van Wyhe and Rookmaaker, *Alfred Russel Wallace: Letters from the Malay Archipelago*, pp. 156–7.
38 Ibid., p. 147.
39 Ibid., p. 212.
40 Wallace, AR, *Island Life: Or the Phenomena and Causes of Insular Faunas and Floras including a Revision and Attempted Solution of the Problem of Geological Climates*, 2nd and rev. edn, Macmillan, London, 1892, p. 375.
41 Wallace, *The Malay Archipelago: The Land of the Orang-Utan, and the Bird of Paradise*, p. 21.
42 National Geospatial-Intelligence Agency, *Sailing Directions (Enroute): Borneo, Java, Sulawesi, and Nusa Tenggara*, Springfield, VA, 2015, p. 127.
43 Wallace, *The Malay Archipelago: The Land of the Orang-Utan, and the Bird of Paradise*, p. 163.
44 National Geospatial-Intelligence Agency, *Sailing Directions (Enroute): Borneo, Java, Sulawesi, and Nusa Tenggara*, pp. 139–40.
45 National Geographic, *Tiny Humans: The Hobbits of Flores*.
46 Morwood, M, and P van Oosterzee, *The Discovery of the Hobbit: the Scientific Breakthrough that Changed the Face of Human History*, Random House, Milsons Point, NSW, 2007, pp. 244–8.
47 Roberts, B, personal communication.
48 Meijer, HJM, et al., 'The Fellowship of the Hobbit: The Fauna Surrounding *Homo floresiensis*', *Journal of Biogeography*, vol. 37, no. 6, 2010, pp. 995–1006; Meijer, HJM, and RA Due, 'A New Species of Giant Marabou Stork (Aves: Ciconiiformes) from the Pleistocene of Liang Bua, Flores (Indonesia)', *Zoological Journal of the Linnean Society*, vol. 160, no. 4, 2010, pp. 707–24. It would have towered over the diminutive *H. floresiensis*.
49 Meijer and Due, 'A New Species of Giant Marabou Stork (Aves: Ciconiiformes) from the Pleistocene of Liang Bua, Flores (Indonesia)'; Van der Geer, A, et al., *Evolution of Island Mammals: Adaptation and Extinction of Placental Mammals on Islands*, Wiley-Blackwell, London, 2010, p. 24; van den Bergh, 'The Late Neogene Elephantoid-Bearing Faunas of Indonesia and Their Palaeozoogeographic Implications'.
50 Meijer et al., 'The Fellowship of the Hobbit: The Fauna Surrounding *Homo floresiensis*'; van den Bergh, GD, et al., 'The Liang Bua Faunal Remains: A 95 K.yr. Sequence from Flores, East Indonesia', *Journal of Human Evolution*, vol. 57, no. 5, 2009, pp. 527–37.
51 Bender, R, and N Bender, 'Brief Communication: Swimming and Diving Behavior in Apes (Pan Troglodytes and Pongo Pygmaeus): First Documented Report', *American Journal of Physical Anthropology*, vol. 152, no. 1, 2013, pp. 156–62.

52 Bird, MI, et al., 'Early Human Settlement of Sahul Was Not an Accident', *Scientific Reports*, vol. 9, 2019, https://www.nature.com/articles/s41598-019-42946-9 (viewed November 2021).

53 Early on in the *H. floresiensis* debate, Maciej Henneberg and colleagues reasoned that had *H. erectus* arrived at Flores once, it could have done so regularly, rather than become isolated and dwarfing (see chapter 2). I also thank an anonymous reader whose thoughts along the same lines I have incorporated here.

6 Big surprise in the Philippines

1 Naumann, E, '*Stegodon mindanensis*, eine Art von Uebergangs-Mastodonten donten', *Abhandlungen und Berichte des Zoologischen und Anthropologisch-Ethnographischen Museums zu Dresden, 1887*, vol. 42, no. 6, 1890, pp. 166–9.

2 Von Koenigswald, GHR, 'Preliminary Report on a Newly Discovered Stone Age Culture from Northern Luzon, Philippines Island', *Asian Perspectives*, vol. 2, no. 2 (Winter), 1958, pp. 69–70.

3 Beyer, HO, *Outline Review of Philippine Archaeology by Islands and Provinces*, Institute of Science, Philippines, 1949, p. 207.

4 Von Koenigswald, 'Preliminary Report on a Newly Discovered Stone Age Culture from Northern Luzon, Philippines Island'.

5 We now know that one of these events occurred 730 000 years ago, when an asteroid or comet impacted South-East Asia. Tektites from this event were strewn over Australia, Indonesia, the Philippines and parts of mainland SE Asia. See Schneider, David A, Dennis V Kent and Gilberto A Mello, 'A Detailed Chronology of the Australasian Impact Event, the Brunhes-Matuyama Geomagnetic Polarity Reversal, and Global Climate Change', *Earth and Planetary Science Letters*, vol. 111, 1992, pp. 395–405.

6 Von Koenigswald, GHR, 'Fossil Mammals from the Philippines', *Proceedings of the Fourth Far-Eastern Prehistory and the Anthropology Division of the Eighth Pacific Science Congresses Combined. Part 1: Archaeology and Physical Anthropology*, p. 359, conference held in Quezon City and Manilla, 16–28 November, 1953.

7 Bautista is a zooarchaeologist at the Archaeological Studies Program. Zooarchaeologists are specialists in animal bones, with expertise in identifying and analysing bones found in archaeological contexts.

8 De Vos, J, personal communication, 22 September 2019.

9 Ibid. For the announcement of this discovery, see Ingicco, T, et al., 'Earliest Known Hominin Activity in the Philippines by 709 Thousand Years Ago', *Nature*, vol. 557, no. 7704, 2018, pp. 233–7.

10 Ingicco, et al., 'Earliest Known Hominin Activity in the Philippines by 709 Thousand Years Ago'.

11 Ronquillo, W, and R Santiago (eds), *Archaeological Cave and Open Sites Exploration at Peneblancă, Cagayan Province*, National Museum of the Philippines, 1977.

12 Cuevas, M (ed.), *Preliminary Report on the Archaeological Excavation Conducted at Callao Cave*, National Museum of the Philippines, 1980.
13 Bellwood, P, International Conference on *Homo luzonensis* and the Hominin Record of Southeast Asia, Quezon City, the Philippines, 3–7 February 2020; Bellwood, P, personal communication, 2019.
14 Mijares, AS, 'Understanding the Callao Cave Depositional History', in HM Philip, J Piper and D Bulbeck (eds), *New Perspectives in Southeast Asian and Pacific Prehistory*, 2017, pp. 125–40.
15 Mijares, AS, et al., 'New Evidence for a 67 000-Year-Old Human Presence at Callao Cave, Luzon, Philippines', *Journal of Human Evolution*, vol. 59, no. 1, 2010, pp. 123–32.
16 Mijares, AS, personal communication, 2020.
17 Mijares, 'Understanding the Callao Cave Depositional History'.
18 Mijares et al., 'New Evidence for a 67 000-Year-Old Human Presence at Callao Cave, Luzon, Philippines'.
19 Philip Piper is now Professor in the ANU School of Archaeology and Anthropology.
20 Piper, P, personal communication, 2019.
21 Ibid.
22 Mijares, AS, personal communication, 2019.
23 Détroit, F, et al., 'Upper Pleistocene *Homo sapiens* from the Tabon Cave (Palawan, the Philippines): Description and Dating of New Discoveries', *Comptes Rendus Palevol*, vol. 3, no. 8, 2004, pp. 705–12.
24 Détroit, F, personal communication, 2019.
25 Zanolli, C, personal communication, 2020.
26 Détroit, F, personal communication, 2019.
27 Ibid.
28 Ibid.
29 Ibid.
30 The Philippine Negritos comprise ethnically distinct groups across the Philippines.
31 Mijares et al., 'New Evidence for a 67 000-Year-Old Human Presence at Callao Cave, Luzon, Philippines'.
32 Ibid.
33 Mijares, AS, Asia-Pacific Conference on Human Evolution, Brisbane, 25–27 June 2019.
34 Mijares, AS, personal communication, 2020.
35 Ibid.
36 Piper, P, personal communication, 2019.
37 Mijares, AS, personal communication, 2019.
38 Détroit, F, personal communication, 2019.
39 Ibid.
40 Détroit, F, et al., 'A New Species of *Homo* from the Late Pleistocene of the Philippines', *Nature*, vol. 568, no. 7751, 2019, pp. 181–6, extended data Figure 1.

41 Or so it seemed to us at the time. Clément Zanolli has since clarified that the team's studies show that the *H. luzonensis* premolars, even if very small, are comparable in size with the lower range of variation of modern humans and close to *H. floresiensis* dimensions: Zanolli, C, personal communication, 2020.
42 Greenaway, J, 'Bones of Contention', *60 Minutes*, 2005.
43 Mijares, AS, personal communication, 2019.
44 No excavations have taken place at Callao Cave since early February 2020, when the novel coronavirus, later named COVID-19, caused a world health emergency.
45 Zanolli, C, personal communication, 2019.
46 Indriati, E, et al., 'The Age of the 20 Meter Solo River Terrace, Java, Indonesia and the Survival of *Homo erectus* in Asia', *Plos One*, vol. 6, no. 6, 2011.
47 Détroit et al., 'A New Species of *Homo* from the Late Pleistocene of the Philippines'.
48 Zanolli, C, personal communication, 2019.
49 Détroit et al., 'A New Species of *Homo* from the Late Pleistocene of the Philippines'.
50 Zanolli, C, personal communication, 2019.
51 Détroit, F, Asia-Pacific Conference on Human Evolution, Brisbane, 25–27 June 2019.
52 Piper, P, personal communication, 2019.
53 Détroit et al., 'A New Species of *Homo* from the Late Pleistocene of the Philippines'.
54 Mijares et al., 'New Evidence for a 67 000-Year-Old Human Presence at Callao Cave, Luzon, Philippines'.
55 Détroit et al., 'A New Species of *Homo* from the Late Pleistocene of the Philippines'.
56 Détroit, F, personal communication, 2020.
57 Détroit et al., 'A New Species of *Homo* from the Late Pleistocene of the Philippines'.
58 From the Koobi Fora region of Africa, dated at 1.5 million years old and referred to as *H. ergaster* or *H. erectus*: see Appendix A.
59 Détroit, Asia-Pacific Conference on Human Evolution.
60 Détroit et al., 'A New Species of *Homo* from the Late Pleistocene of the Philippines'.
61 Zanolli, C, personal communication, 2019.
62 Mijares, AS, personal communication, 2020.
63 Ibid.
64 Slon, V, et al., 'Neanderthal and Denisovan DNA from Pleistocene Sediments', *Science*, vol. 356, no. 6338, 2017, pp. 605–8.
65 Mijares, Asia-Pacific Conference on Human Evolution, Brisbane, 25–27 June 2019.
66 Détroit et al., 'A New Species of *Homo* from the Late Pleistocene of the Philippines'.

67 Piper, P, personal communication, 2020.
68 Ibid.
69 Ibid.
70 Mijares, AS, personal communication, 2020.

Reactions to *Homo luzonensis*

1 Détroit, F, Asia-Pacific Conference on Human Evolution, Brisbane, 25–27 June 2019.
2 Détroit, F, personal communication, 2019.
3 Mijares, AS, '*Homo luzonensis*: A New Species of Hominin from the Late Pleistocene of Northern Luzon, Philippines', speakers Professor Armand Mijares, University of the Philippines; and Professor Philip Piper, Australian National University, public presentation at the Australian National University, Canberra, hosted by ANU College of Arts & Social Sciences, and sponsored by the Embassy of the Philippines 1 July 2019.

Let's go digging

1 Stahlschmidt, MC, et al., 'Ancient Mammalian and Plant DNA from Late Quaternary Stalagmite Layers at Solkota Cave, Georgia', *Scientific Reports*, Vol 9, no. 6628, 2019, pp. 1-10.

Appendix A Cousins by the dozens

1 Langergraber, KE, et al., 'Generation Times in Wild Chimpanzees and Gorillas Suggest Earlier Divergence Times in Great Ape and Human Evolution', *Proceedings of the National Academy of Sciences of the United States of America*, vol. 109, no. 39, 2012, pp. 15716–21.
2 And possibly from an isolated upper fourth premolar (P4) from Azmaka in Bulgaria: Fuss, J, et al., 'Potential Hominin Affinities of Graecopithecus from the Late Miocene of Europe', *Plos One*, vol. 12, no. 5, 2017.
3 Bohme, M, et al., 'Messinian Age and Savannah Environment of the Possible Hominin Graecopithecus from Europe', *Plos One*, vol. 12, no. 5, 2017.
4 Technically, the lower fourth premolar (P4): Fuss et al., 'Potential Hominin Affinities of Graecopithecus from the Late Miocene of Europe'.
5 Ibid.
6 Senut, B, et al., 'First Hominid from the Miocene (Lukeino Formation, Kenya)', *Comptes Rendus De L Academie Des Sciences Serie Ii Fascicule a-Sciences De La Terre Et Des Planetes*, vol. 332, no. 2, 2001, pp. 137–44.
7 Brunet, M, et al., 'A New Hominid from the Upper Miocene of Chad, Central Africa', *Nature*, vol. 418, no. 6894, 2002, pp. 145–51, erratum in vol. 418, no. 6899, 2002, p. 801.
8 White, TD, G Suwa and B Asfaw, '*Australopithecus ramidus*, a New Species of Early Hominid from Aramis, Ethiopia', *Nature*, vol. 371, no. 6495, 1994, pp. 306–12; White, TD, et al., '*Ardipithecus ramidus* and the Paleobiology of Early Hominids',

Science, vol. 326, no. 5949, 2009, pp. 75–86; Wood, B, and E Boyle, 'Hominins: Context, Origins, and Taxic Diversity', in Michel Tibayrenc and Francisco J Ayala (eds), *On Human Nature: Biology, Psychology, Ethics, Politics, and Religion*, Academic Press, Cambridge, MA, 2017, pp. 17–44; Haile-Selassie, Y, G Suwa and TD White, 'Late Miocene Teeth from Middle Awash, Ethiopia, and Early Hominid Dental Evolution', *Science*, vol. 303, no. 5663, 2004, pp. 1503–5.

9 Georgiou, L, et al., 'Evidence for Habitual Climbing in a Pleistocene Hominin in South Africa', *Proceedings of the National Academy of Sciences of the United States of America*, vol. 117, no. 15, 2020, pp. 8416–23.

10 The other four are *A. deyiremeda*, *A. anamensis*, *A. bahrelghazali* and *A. prometheus*.

11 Wood and Boyle, 'Hominins: Context, Origins, and Taxic Diversity'.

12 Smithsonian National Museum of National History, '*Australopithecus afarensis*', January 2021, https://humanorigins.si.edu/evidence/human-fossils/species/australopithecus-afarensis (viewed November 2021).

13 Schoenemann, T, 'Hominid Brain Evolution', in DR Begun (ed.), *A Companion to Paleoanthropology*, Blackwell Publishing, London, 2013, pp. 136–64.

14 Wood and Boyle, 'Hominins: Context, Origins, and Taxic Diversity'.

15 Schoenemann, 'Hominid Brain Evolution'.

16 Smithsonian National Museum of National History, '*Australopithecus afarensis*'. This difference between male and females is called sexual dimorphism.

17 Asfaw, B, et al., '*Australopithecus garhi*: A New Species of Early Hominid from Ethiopia', *Science*, vol. 284, no. 5414, 1999, pp. 629–35.

18 Berger, LR, '*Australopithecus sediba*: A New Species of *Homo*-like Australopith from South Africa', vol. 328, no. 5975, 2010, pp. 195–204.

19 Ibid.

20 Ibid.

21 DeSilva, JM, et al., 'The Lower Limb and Mechanics of Walking in *Australopithecus sediba*', *Science*, vol. 340, no. 6129, 2013.

22 Rein, TR, et al., 'Adaptation to Suspensory Locomotion in *Australopithecus sediba*', *Journal of Human Evolution*, vol. 104, 2017, pp. 1–12.

23 Wood and Boyle, 'Hominins: Context, Origins, and Taxic Diversity'.

24 Schoenemann, 'Hominid Brain Evolution'.

25 Richmond, BG, et al., 'The Upper Limb of *Paranthropus boisei* from Ileret, Kenya', *Journal of Human Evolution*, vol. 141, 2020, pp. 1–21.

26 For example, see McPherron, Shannon P, et al., 'Evidence for Stone-Tool-Assisted Consumption of Animal Tissues before 3.39 Million Years Ago at Dikika, Ethiopia', *Nature*, vol. 466, 2010, pp. 857–60.

27 Villmoare, B, et al., 'Early *Homo* at 2.8 Ma from Ledi-Geraru, Afar, Ethiopia', *Science*, vol. 347, no. 6228, 2015, pp. 1352–5.

28 Wood and Boyle, 'Hominins: Context, Origins, and Taxic Diversity'.

29 Schoenemann, 'Hominid Brain Evolution'.

30 Leakey, LSB, JR Napier and PV Tobias, 'New Species of Genus *Homo* from Olduvai Gorge', *Nature*, vol. 202, no. 492s, 1964, pp. 5–7.

31 Leakey, REF, 'Evidence for an Advanced Plio-Pleistocene Hominid from East Rudolf, Kenya', *Nature*, vol. 242, no. 5398, 1973, pp. 447–50.
32 Some researchers view one of the partial leg bones as non-hominin.
33 Herries, AIR, et al., 'Contemporaneity of *Australopithecus*, *Paranthropus*, and early *Homo erectus* in South Africa', *Science*, vol. 368, no. 47, 2020.
34 Swisher, CC, et al., 'Age of the Earliest Known Hominids in Java, Indonesia', *Science*, vol. 263, no. 5150, 1994, pp. 1118–21.
35 Larick, R, et al., 'Early Pleistocene Ar-40/Ar-39 Ages for Bapang Formation Hominins, Central Java, Indonesia', *Proceedings of the National Academy of Sciences of the United States of America*, vol. 98, no. 9, 2001, pp. 4866–71.
36 Matsu'ura, S, et al., 'Age Control of the First Appearance Datum for Javanese *Homo Erectus* in the Sangiran Area', *Science*, vol. 367, no. 6474, 2020, pp. 210–14.
37 Hyodo, M, et al., 'High-Resolution Record of the Matuyama-Brunhes Transition Constrains the Age of Javanese *Homo erectus* in the Sangiran Dome, Indonesia', *Proceedings of the National Academy of Sciences of the United States of America*, vol. 108, no. 49, 2011, pp. 19563–8.
38 Grun, R, et al., 'ESR Analysis of Teeth from the Paleoanthropological Site of Zhoukoudian, China', *Journal of Human Evolution*, vol. 32, no. 1, 1997, pp. 83–91.
39 Schoenemann, 'Hominid Brain Evolution'.
40 Groves, CP, and V Mazák, 'An Approach to the Taxonomy of the Hominidae: Gracile Villafranchian Hominids of Africa', *Casopis pro mineralogii a geologii*, vol. 20, 1975, pp. 225–46.
41 Larson, SG, 'Evolutionary Transformation of the Hominin Shoulder', *Evolutionary Anthropology*, vol. 16, no. 5, 2007, pp. 172–87.
42 Meikle, WE, and ST Parker, *Naming Our Ancestors: An Anthology of Hominid Taxonomy*, Waveland Press, Prospect Heights, IL, 1994, p. 120.
43 Groves and Mazák, 'An Approach to the Taxonomy of the Hominidae: Gracile Villafranchian Hominids of Africa'.
44 Gabounia, L, MA de Lumley, A Vekua, A Lordkipanidze and H de Lumley, 'Découverte d'un nouvel hominidé à Dmanissi (Transcaucasie, Géorgie)', *Comptes Rendus Palevol*, vol. 1, no. 4, 2002, pp. 243–353.
45 Garcia, T, et al., 'Earliest Human Remains in Eurasia: New 40Ar/ 39Ar Dating of the Dmanisi Hominid-Bearing Levels, Georgia', *Quaternary Geochronology*, vol. 5, no. 4, 2010, pp. 443–51.
46 Groves, CP, *A Theory of Human and Primate Evolution*, Clarendon Press, Oxford, 1989, p. 4.
47 Alex, B, 'Meet the Denisovans', *Discover Magazine*, 4 November 2016.
48 When Professor Derevianko visited the ANU in February 2012, he gently mentioned to a group of us over lunch that Denisova is pronounced *den-ee-sova*. We had been pronouncing it with a distinct Australian accent—*deni-sova*—and of course had it quite wrong.

49 Krause, J, et al., 'The Complete Mitochondrial DNA Genome of an Unknown Hominin from Southern Siberia', *Nature*, vol. 464, no. 7290, 2010, pp. 894–7, information on p. 894.
50 Reich, D, et al., 'Genetic History of an Archaic Hominin Group from Denisova Cave in Siberia', *Nature*, vol. 468, no. 7327, 2010, pp. 1053–60.
51 Krause, 'The Complete Mitochondrial DNA Genome of an Unknown Hominin from Southern Siberia', p. 894.
52 Ibid.; Meyer, M, et al., 'Nuclear DNA Sequences from the Middle Pleistocene Sima de Los Huesos Hominins', *Nature*, vol. 531, no. 7595, 2016, pp. 504–7, information on p. 504.
53 Reich et al., 'Genetic History of an Archaic Hominin Group from Denisova Cave in Siberia'; Slon, V, et al., 'The Genome of the Offspring of a Neanderthal Mother and a Denisovan Father', *Nature*, vol. 561, no. 7721, 2018, pp. 113–16.
54 Slon et al., 'The Genome of the Offspring of a Neanderthal Mother and a Denisovan Father'. See also Brown, S, et al., 'Identification of a New Hominin Bone from Denisova Cave, Siberia Using Collagen Fingerprinting and Mitochondrial DNA Analysis', *Scientific Reports*, vol. 6, 2016.
55 The Leakey Foundation, 'Neanderthal Mother, Denisovan Father', 23 August 2018, https://leakeyfoundation.org/neanderthal-mother-denisovan-father (viewed November 2021).
56 Slon, V, et al., 'The Genome of the Offspring of a Neanderthal Mother and a Denisovan Father', *Nature*, vol. 561, no. 7721, 2018, pp. 113–16.
57 Zimmer, C, 'Denisova Jawbone Discovered in a Cave in Tibet', *The New York Times*, 1 May 2019.
58 Welker, F, 'Denisovans in Tibet', Nature Portfolio Ecology and Evolution Community, 2 May 2019, https://go.nature.com/2DMNZZE (viewed November 2021).
59 Chen, FH, et al., 'A Late Middle Pleistocene Denisovan Mandible from the Tibetan Plateau', *Nature*, vol. 569, no. 7756, 2019, pp. 409–12.
60 Ibid.
61 Huerta-Sanchez, E., et al, 'Altitude Adaptation in Tibetans Caused by Introgression of Denisovan-Like DNA', *Nature*, vol. 512, no. 7513, 2014, pp. 194–7.
62 Berger, LR, et al., '*Homo naledi*, a New Species of the Genus *Homo* from the Dinaledi Chamber, South Africa', *Elife*, vol. 4, 2015.
63 Robbins, JL, et al., 'Providing Context to the *Homo naledi* Fossils: Constraints from Flowstones on the Age of Sediment Deposits in Rising Star Cave, South Africa', *Chemical Geology*, vol. 567, 2021.
64 Berger et al., '*Homo naledi*, a New Species of the Genus *Homo* from the Dinaledi Chamber, South Africa'.
65 Schoenemann, 'Hominid Brain Evolution'.
66 Feuerriegel, EM, et al., 'The Upper Limb of *Homo naledi*', *Journal of Human Evolution*, vol. 104, 2017, pp. 155–73; Feuerriegel, EM, et al., 'The Shoulder

and Upper Limb of *Homo naledi*', *American Journal of Physical Anthropology*, vol. 159, 2016, p. 142.

67 Kivell, TL, et al., 'The Hand of *Homo naledi*', *Nature Communications*, vol. 6, 2015.

68 Berger et al., '*Homo naledi*, a New Species of the Genus *Homo* from the Dinaledi Chamber, South Africa'.

69 Hooper, R, '*Homo naledi*: Unanswered Questions about the Newest Human Species', *New Scientist*, 10 September 2015.

70 Hublin, JJ, et al., 'New Fossils from Jebel Irhoud, Morocco and the Pan-African Origin of *Homo sapiens*', *Nature*, vol. 546, 8 June 2017, pp. 289–92, correction in vol. 558, 13 June 2018, p. E6.

Appendix B Hominin fossil discoveries

1 Often just called Neanderthals.

2 Shipman, P, *The Man Who Found the Missing Link*, Simon and Schuster, New York, 2001, p. 56.

3 Devos, J, and F Aziz, 'The Excavations by Dubois (1891–1900), Selenka (1906–1908), and the Geological Survey by the Indonesian-Japanese Team (1976–1977) at Trinil (Java, Indonesia)', *Journal of the Anthropological Society of Nippon*, vol. 97, no. 3, 1989, pp. 407–20, information on p. 407.

4 Shipman, *The Man Who Found the Missing Link*, p. 95.

5 Devos and Aziz, 'The Excavations by Dubois (1891–1900), Selenka (1906–1908), and the Geological Survey by the Indonesian-Japanese Team (1976–1977) at Trinil (Java, Indonesia)', p. 408.

6 Ibid.

7 Shipman, *The Man Who Found the Missing Link*, pp. 141–3.

8 Meikle, WE, and ST Parker, *Naming Our Ancestors: An Anthology of Hominid Taxonomy*, Waveland Press, Prospect Heights, IL, 1994, pp. 37–40.

9 Ibid., pp. 36–7.

10 Dart, RD, and D Craig, *Adventures of the Missing Link*, Hamish Hamilton, London, 1959.

11 Ibid., p. 3.

12 Ibid., p. 4.

13 Ibid., p. 6.

14 Ibid., p. 8.

15 There is a wonderful photo of the Taung skull nesting comfortably within the palm of a man's hand (Dart's?) in GHR von Koenigswald's *Meeting Prehistoric Man* (Harper & Brothers, New York, 1956, opposite p. 161).

16 Dart and Craig, *Adventures of the Missing Link*, p. 9.

17 Miss Goodall is the famous primatologist Dr Jane Goodall.

18 Leakey, LSB, 'Recent Discoveries at Olduvai Gorge, Tanganyika', *Nature*, vol. 181, no. 4616, 1958, pp. 1099–103.

19 Leakey, LSB, 'New Finds at Olduvai Gorge', *Nature*, vol. 189, no. 476, 1961, pp. 649–50.
20 Now known as Olduvai Hominid 12 (OH 12).
21 Leakey, LSB, and MD Leakey, 'Recent Discoveries of Fossil Hominids in Tanganyika: At Olduvai and Near Lake Natron', *Nature*, vol. 202, 1964, pp. 5–7.
22 Leakey, LSB, JR Napier and PV Tobias, 'New Species of Genus *Homo* from Olduvai Gorge', *Nature*, vol. 202, 1964, pp. 7–9. John Napier was also an orthopaedic surgeon and palaeoanthropologist.
23 Leakey, M, *Disclosing the Past*, The Rainbird Publishing Group, London, 1984, pp. 120–1.
24 Ibid., pp. 124–5.
25 Johanson, DC, and MA Edey, *Lucy: The Beginnings of Humankind*, Simon and Schuster, New York, 1981, p. 16.
26 Ibid., p. 18.
27 Wong, K, 'Stranger in a New Land', *Scientific American Special Editions*, vol. 16, 2006, pp. 38–47.
28 Gabunia, L, and A Vekua, 'A Pliopleistocene Hominid from Dmanisi, East Georgia, Caucasus', *Nature*, vol. 373, no. 6514, 1995, pp. 509–12.
29 Wong, 'Stranger in a New Land'.
30 Balter, M, and A Gibbons, 'A Glimpse of Humans' First Journey out of Africa', *Science*, vol. 288, no. 5468, 2000, pp. 948–50.
31 *Malapa* means 'homestead' in the Sesotho language: see Berger, LR, et al., '*Australopithecus sediba*: A New Species of *Homo*-Like Australopith from South Africa', *Science*, vol. 328, no. 5975, 2010, pp. 195–204.
32 'The Discovery of *Australopithecus sediba*', YouTube, 25 November 2020, https://www.youtube.com/watch?v=5YEiJVQdI-Q (viewed November 2021).
33 Wong, Kate, 'Mystery Human', *Scientific American*, vol. 314, no. 3, March 2016, pp. 20–30. Also see Brahic, Catherine, 'Chamber of Secrets', *New Scientist*, vol. 2997, 29 November 2014, pp. 41–3.
34 Feuerriegel, EM, personal communication, 5 January 2020.
35 Hawks, J, et al., 'New Fossil Remains of *Homo naledi* from the Lesedi Chamber, South Africa', *Elife Sciences*, vol. 6, 2017, https://elifesciences.org/articles/24232 (viewed November 2021).

Glossary

1 Hine, Robert (ed.), *Dictionary of Biology*, 8th edn, Oxford University Press, Oxford, 2019. Also see Beth Blaxland, 'Hominid and Hominin: What's the Difference?', Australian Museum, 2 October 2020, https://australian.museum/learn/science/human-evolution/hominid-and-hominin-whats-the-difference (viewed November 2021).
2 *Hutchinsons Dictionary of Science*, ebook, Helicon Publishing, Boston, 1994.

3 Ibid.
4 Gibbard, Philip L, Martin J Head and Michael JC Walker, 'The Subcommittee on Quaternary Stratigraphy, 2010: Formal Ratification of the Quaternary System/Period and the Pleistocene Series/Epoch with a Base at 2.58 MA', *Journal of Quaternary Science*, February 2010, pp. 96–102.

Acknowledgements

I am grateful to my much-missed colleague, Emeritus Professor Colin Groves. Colin and I worked together on the *H. floresiensis* question from the time of its announcement until Colin's passing. He was a true scholar and an inspiring teacher. I am deeply indebted to Colin for his collegiality, his clear and objective thinking, his intellectual acumen, his sense of humour and his unflappable nature.

I thank the late Professor Mike Morwood, Dr Tony Djubiantono, Dr Thomas Sutikna and ARKENAS for kindly affording me the immense privilege of studying the *H. floresiensis* bones—a highlight of my life. Mike and Thomas I also thank for facilitating my visits to Flores and Liang Bua cave. A thank you, too, to members of the Liang Bua team for your support of my research endeavours on Flores.

It was a pleasure to work with Richard Wright, Denise Donlon, Dave Cameron, Bill Jungers and Mike Lee on the *H. floresiensis* question. Thank you all for your collegiality and contributing your expertise.

Many people generously shared personal recollections of their discoveries and research work with me for inclusion in this book. Your experiences have greatly enriched this work and I am grateful to each of you for your willingness to share your insights, and for taking time out of your busy days to chat with me. Thank you Linda Ayliffe, Peter Bellwood, Adam Brumm,

David Cameron, Jon de Vos, Florent Détroit, Elen Feuerriegel, Mike Gagan, Wahyoe Hantoro, Jatmiko, Susan Larson, Mike Lee, Glenn Marshall, Hanneke Meijer, Armand (Mandy) Mijares, Phil Piper, Bert Roberts, Heather Scott-Gagan, Garry K Smith, Thomas Sutikna, Matt Tocheri, Gert van den Bergh, Frido Welker and Clément Zanolli.

Other colleagues and friends contributed feedback and insights, or helped in other ways: Hannah Bulloch, Christine Cave, Marilyn Chalkley, Phyll Dance, Rokus Awe Due, Dean Falk, Stewart Fallon, Natasha Fijn, Gregory Forth, Heloisa Mariath, Michael McFadden, Mark Moore, Anton Nurcahyo, Mark Oxenham, Myron Shekelle and Lawrie St Hill.

Very early in the piece, Julie Jefferis made a most generous offer to read my draft chapters. This is a time-consuming task that requires patience and care. Thank you Julie for your thoughtful reading and for your insightful suggestions and ongoing encouragement. These were of invaluable help along the way.

Geraldine Cave crafted the maps and diagrams beautifully. Thank you Geraldine for your willingness to do this, and for your ingenuity in solving the pictorial challenges inherent in presenting information diagrammatically.

I thank Andrew Schuller, who considered my book to have publishing potential and kindly took up the case.

Nathan Hollier, Duncan Fardon, Cathryn Smith and the team at Melbourne University Publishing have been great to work with. I very much value the input of copy-editor Paul Smitz. I imagine working under COVID conditions would have been very challenging for everyone and I appreciate how nicely things flowed.

I thank Stephen Oppenheimer and several anonymous reviewers for their helpful comments on earlier drafts of various chapters.

I sincerely thank an anonymous reader of the book who provided useful comments as well as some extra information. I hope I have done justice to the remarks, and I have incorporated the new information, duly attributed to the reader.

The Australian National University, the School of Archaeology and Anthropology, and the College of Arts and Social Sciences have supported me throughout my studies. I very much appreciate the opportunities this has afforded me and I truly enjoy being part of such a lively academic community.

The Australian Research Council has been most generous in providing the financial backing for Colin Groves' and my *H. floresiensis* research project DP1096870.

Very special thanks go to my husband, Fraser Argue, for his boundless support and enthusiasm for my research.

I will be mortified if I have inadvertently missed someone in my acknowledgements. If this has happened, I am deeply apologetic.

Index